Introduction to Computational Geometry

INTRODUCTION TO COMPUTATIONAL GEOMETRY

Prof. Avanish Kumar

ARCLER
PRESS

www.arclerpress.com

Introduction to Computational Geometry

Prof. Avanish Kumar

Arcler Press
224 Shoreacres Road
Burlington, ON L7L 2H2
Canada
www.arclerpress.com
Email: orders@arclereducation.com

ISBN: 978-1-77469-847-1 (Hardcover)

ABOUT THE AUTHOR

Prof. Avanish Kumar is presently working as Professor at the Department of Mathematical Sciences and Computer Applications, Bundelkhand University, Jhansi. He has completed his M.Sc., from Meerut University (presently CCS University), Meerut in 1989, M. Phil., from University of Roorkee (presently IIT), Roorkee in 1992 and Ph. D. in Mathematics, from Gurukula Kangri, University, Hardwar in 1999. University of Roorkee awarded him 'University Medal' for standing First in First Class. He has served as Assistant Scientific Officer in the Commission for Scientific and Technical Terminology, Ministry of Human Resource Development, Government of India, New Delhi w.e.f. 20 November 1995 to 25 November 1999. He joined Bundelkhand University as Assistant Professor in the Department of Mathematical Sciences and Computer Applications on 26 November 1999. There he served in various capacity and became Professor on 26 November 2014. He served as Chairman of the Commission for Scientific and Technical Terminology, Department of Higher Education, Ministry of Education w.e.f. 06 February 2017 to 05 February 2021. He was having additional charge of the Director, Central Hindi Directorate, Department of Higher Education, Ministry of Education, Government of India w. e. f. 07 July 2017 to April 08, 2020. He has supervised 14 Ph.D. thesis, 28 M. Phil. Dissertation. He has published about 90 research papers in National and International Journal of reputes. He is the author of two books and editor of four terminological related books. His research interests are Modeling, Optimization Techniques and Distributed Computing Systems. He is the life member of International Academy of Physical Sciences, Indian Mathematical Society, Computer Society of India, Operations Research Society of India, and many other similar academic bodies and societies. Also, he is the Vice President, of Operations Research Society of India (ORSI) and Vijñāna Parishad of India (VPI). During 2010 to 2013 he was the Head of the Department of Mathematical Sciences and Computer Applications. During the same period 2010 to 2013 he was also the Head of the Institute of Computer Science and Information Technology. He was the Chief of University Employment Information and Counseling Centre (UP Government) Bundelkhand University, Jhansi during 11 June 2009 to 05 February 2017. During this period Ministry of Employment and Labor, Government of India, New Delhi awarded the establishment of Modal Career Centre (MCC) under the University Employment Information and Counseling Centre (UP Government) Bundelkhand University, Jhansi to facilitate the job seekers of this university as well as Bundelkhand Region with career related trainings, guidance and counseling. He was visited to the Rajamangla University of Technology. Thanyaburi (Thailand), University of Science and Culture, Tehran (Iran), Kathmandu, (Nepal), Port Louise (Mauritius), Shanghai (China), Guadalajara (Mexico) for various academic work. In 2000, he was awarded 'Professor R. C. Mehrotra Best Teacher Award and Gold Medal' for outstanding contribution to the university services by the Bundelkhand University, Jhansi and in 2016, Bharat Vikas Award was conferred to him by the Institute of Self Reliance, Bhubaneswar, Odisha, in appreciation of dedicated services to the society and humanity.

TABLE OF CONTENTS

LIST OF FIGURES

LIST OF TABLES

LIST OF ABBREVIATIONS

BSP	binary space partitioning
CAM	computer-aided manufacturing
DOF	degrees of freedom
DTM	digital terrain model
GIS	geographic information systems
HCI	human-computer interaction
RSG	regular square grid
TINs	triangulated irregular networks

PREFACE

Computing is typically depicted as a high-tech field that advances at a rapid pace determined by Moore's Law. This is not entirely accurate. We run the danger of missing a game-changing technological innovation or a paradigm-shifting theoretical discovery if we take our eyes off the ball for a few moments. This book takes a different approach, presenting computers as a science with generally applicable rules that can be applied to any situation. The study of data processing is known as computer science. A new vocabulary is required to explain science, and this book presents the framework of large principles as a starting point for developing such a vocabulary. This book is on computing as a whole, including its algorithms, architectures, and designs, among other things.

Algorithms are the primary subject of this book on introduction to computational geometry. All techniques are tied to specific applications in the fields of robotics, graphics, CAD/CAM, and geographic information systems (GIS), which provides motivation for their development. Modern insights in computational geometry are applied in order to give solutions that are both efficient and simple to grasp and put into practice.

In the late 1970s, computational geometry arose from the field of algorithm creation and analysis. It has evolved into a recognized subject, complete with publications, conferences, and a significant community of active scholars. The beauty of the problems studied and the solutions obtained, on the one hand, can explain the field's success as a research discipline, and on the other hand, the many application domains—computer graphics, geographic information systems (GIS), robotics, and others—in which geometric algorithms play a fundamental role.

Early algorithmic solutions to many geometric problems were either slow or difficult to understand and apply. A variety of new algorithmic strategies have been created in recent years that have improved and simplified many of the prior approaches. We attempted to make these new algorithmic solutions accessible to a wide audience in this textbook. The book was designed as a textbook for a computational geometry course, but it can also be used for self-study.

Except for the introductory chapter, each of the sixteen chapters begins with a problem that arises in one of the application domains. This problem is then turned into a purely geometric one, which is solved using computational geometry techniques. Each chapter's real topic is the geometry issue and the concepts and techniques required to solve it. The applications were chosen based on the concepts in computational geometry that we wished to cover; they are not intended to provide comprehensive coverage of the application domains. The applications' goal is to excite the reader; the chapters' goal is not to give ready-to-use solutions for them. Having stated that, we feel that understanding of computational geometry is essential for efficiently solving geometric problems in application fields. We believe that our book will pique the interest of people not only in the algorithms community, but also in the application domains. Even when multiple solutions exist for most geometric problems, we provide only one solution. In general, we choose the option that is the simplest to comprehend and apply. This isn't always the most efficient solution. We also made certain that the book had a nice mix of techniques such as divide-and-conquer, plane sweep, and randomized algorithms. We decided not to cover all possible modifications to the problems; instead, we felt it was more necessary to cover the major subjects in computational geometry.

This book is mandatory reading for professionals in science and engineering fields that include a "computational" component, for computing practitioners looking for overviews of less familiar areas of computer science, and for non-computer science majors looking for an accessible introduction to the field.

—***Author***

CHAPTER

1 Fundamentals of Computational Geometry

Contents

1.1. INTRODUCTION

The electronic computer was one of the most significant advances of the twentieth century. The computer, and the communication and information technologies built on it, have changed business, government, society, and science, and have had an impact on almost every aspect of our lives, just as industrialization did in the nineteenth century. The main principles and techniques used in the growth of computer apps are covered in this literature, which covers the subject of computing.

Getting into a new industry, such as computers, is akin to visiting a workstation in a foreign country for the first time. While all countries have certain essential qualities, such as the requirement for language and cultural and commerce preferences, the significant differences in such traits from one country to the next can be confusing and even disastrous for newcomers. Furthermore, identifying a country's features in a general way is difficult because they vary greatly and change over time. Similarly, entering the world of computers can be disconcerting, and defining its qualities can be difficult (Figure 1.1).

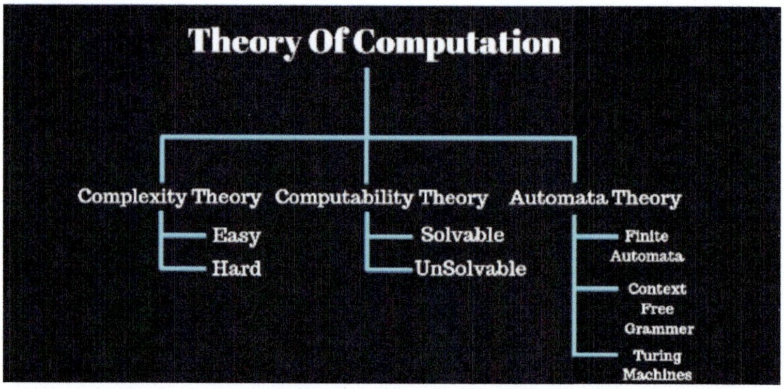

Figure 1.1. The computation theories.

Source: https://www.proprofs.com/quiz-school/story.php?title=njuyodkw2z8j.

Although there are basic notions that underpin the subject of computers which may be defined, learnt, and applied practically. All calculation is dependent upon the associated utilization of computer types of equipment, referred to as hardware, and the programs of computers that make them referred to as software. All the software apps have been designed utilizing data and the specifications of the process, referred to as algorithms and data structures, and all hardware devices are manufactured by utilizing algorithms and data structures. Although software and hardware technology have advanced continuously during the history of computing, and novel paradigms for process and data descriptions have emerged regularly, such foundations have maintained relatively consistent throughout that period.

It begins with defining the idea of computing, and then moves on to examine the notions of software and hardware, before concluding with a foreword to the creation of software (also known as the programming of the computer). After that, most work is devoted to the production of computer software, with an in-depth explanation of software concepts and a glimpse of contemporary culture in the software development area. For the first half of the work, the author makes extensive usage of Processing, a Java-based programming environment; after that, the author makes extensive usage of the complete Java development environment.

1.2. COMPUTING

As previously said, defining computing is difficult, however, according to the curriculum of computing 2005: The Overview Report issued through a joint committee of AIS, ACM, and IEEE, computing is defined as follows: "We may define computing in a broad sense as any goal-oriented activity that needs, advantages from, or creates computers.

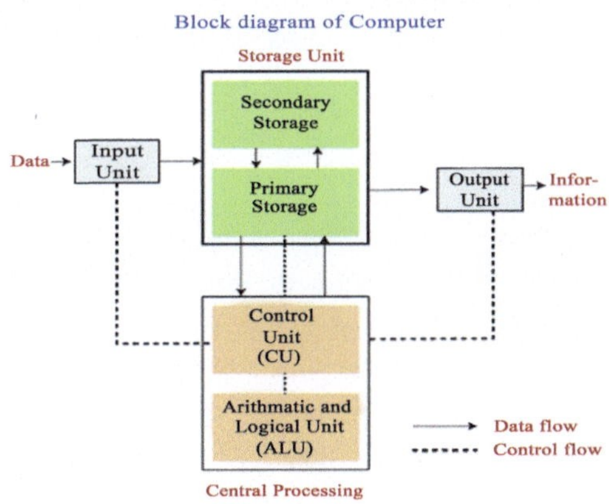

Figure 1.2. The computer's block diagram.

Source: https://www.tutorialandexample.com/block-diagram-of-a-computer/.

It is a wide idea that contains computer hardware creation, the development of the application, and the development of software. The production of the software of a computer is the latest of such businesses and the subject of this text (Figure 1.2) (Agrawal et al., 2006). Since the development of software is reliant on computer hardware, we will cover the basics of computer hardware and how it relates to software to better get ready students for the construction of software. The authors believe that it would connect a fresh era of software engineers, such as not just scientifically and mathematically inclined students who seem to be prevalent in the courses of programming, as well as a younger youth of students in the social sciences and civilizations that have been discovering that calculation is as pertinent to their disciplines as it is always in the fields of science.

Consider the following scenario: you are wandering around a university campus when you suddenly remember you need to make an important phone call. There have been a lot of public phones on campus, and you would like to use the one that is closest to you. It will be beneficial with a campus map upon which you might locate the nearest public phone, regardless of where you were on campus. The map must depict a partition of the campus into areas, with the location of the closest public mobile phone for every region indicated (Posdamer, 1977). What might the landscape look like in such areas? And how would we go about calculating them?

Although this is a little point, it explains the fundamentals of a geometric notion that is used in a variety of applications. The campus is divided into sections using a Voronoi diagram. It may be utilized to represent diverse city trading regions, direct robots, and even define and simulate crystal development. Geometric algorithms are required to compute a geometric structure such as a Voronoi figure. This book's focus is on such algorithms (Noltemeier, 1988; Ghosh, 2000).

Here's another instance. Suppose you've found the nearest public telephone. Having a map of campus in hand, getting to the phone along a pretty shorter pathway with no striking walls or other obstacles should be rather easy. However, training a robot to do the same thing is far more complex. The challenge is geometric again: given a set of geometric barriers, we must identify a short path between two places while preventing collisions with the obstacles. In robotics, resolving the so-called motion planning issue is critical (Gu et al., 2010).

The 3rd case in point: Suppose you have two maps: first with an explanation of the different buildings, containing public mobile phones, and second with a map of the campus's highways. To arrange a switch to a public mobile phone, we must superimpose such maps that merge the data from the two maps. One of the most fundamental functions of geographic information systems (GIS) is to overlay maps. It entails finding things from one map in another, calculating the intersection of different characteristics, etc. (Goodrich et al., 1993).

These are only three instances of geometric issues that need the use of well-designed geometric algorithms to solve. The area of computational geometry, which deals with these geometric issues, arose in the 1970s. It is the systematic examination of structures of data and algorithms for geometric objects, having a focal point upon asymptotically fast accurate algorithms (Sethian, 1999). The challenges provided by geometric issues drew a large number of scholars in.

The path from issue conceptualization to an effective and excellent solution is long and winding, with several challenging and sub-optimal intermediate outcomes. Nowadays, there has a large number of effective geometric algorithms which are also reasonably simple to learn and execute (Figure 1.3) (Held, 1991).

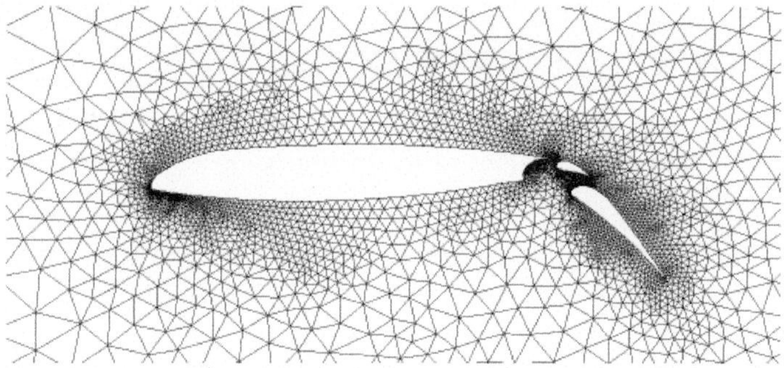

Figure 1.3. Example of a computational model.

Source: https://www.ics.uci.edu/~eppstein/161/960307.html.

This book presents the most significant concepts, methods, structures of data and algorithms in calculational geometry in a style that would appeal to readers interested in implementing computational geometry results. Every chapter is inspired by a real-world situation that necessitates the use of geometric algorithms to solve.

The challenges were chosen from a variety of application domains to demonstrate the extensive applicability of computational geometry: graphics of computer, robotics, CAM /CAD, and GIS (Figure 1.4) (De Floriani et al., 2000).

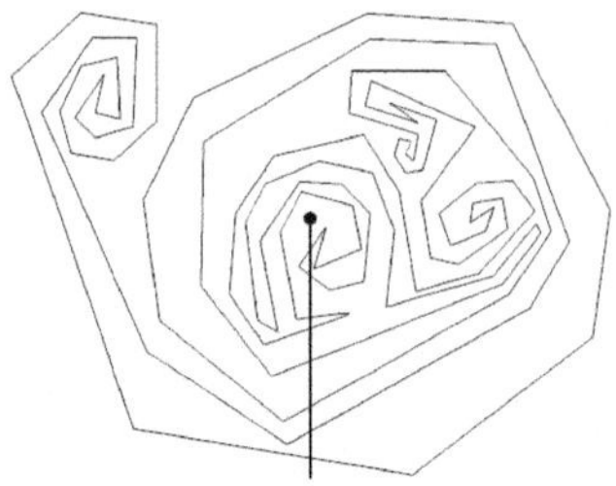

Figure 1.4. Testing if a point is in a polygon.

Source: https://www.ics.uci.edu/~eppstein/161/960307.html.

For serious challenges in the application fields, don't anticipate ready-to-use software solutions. Each chapter focuses on a particular computational geometry idea, with the applications serving simply to motivate and introduce the principles. They also show how to model engineering issues and come up with a precise answer (De Berg et al., 2008).

1.3. AN EXAMPLE: CONVEX HULLS

The majority of best solutions to geometric algorithmic issues are built on two elements. The first is a deep grasp of the issue's geometric aspects, and the second is the correct use of algorithmic approaches and data structures. Every algorithm worldwide did not assist in resolving the issue efficiently if we don't comprehend the issue's geometry. On either side, even though you completely comprehend the geometry of the issue, if someone doesn't understand the correct tools of algorithmic, it will be difficult to resolve it efficiently. We will teach you how to use the most significant geometric ideas and computational procedures (Figure 1.5) (Preparata et al., 1985).

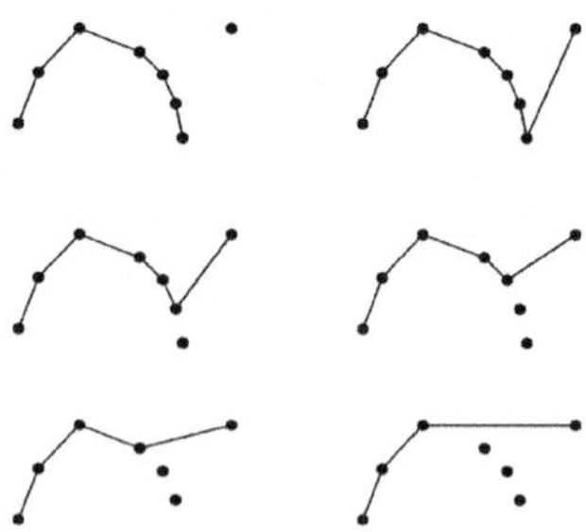

Figure 1.5. Representation of convex hulls.

Source: https://www.ics.uci.edu/~eppstein/161/960307.html.

This section deals with one of the earliest topics explored in calculational geometry: the calculation of planar convex hulls, to highlight the complications which occur while designing a geometric algorithm (Figure 1.6) (Barber et al., 1996).

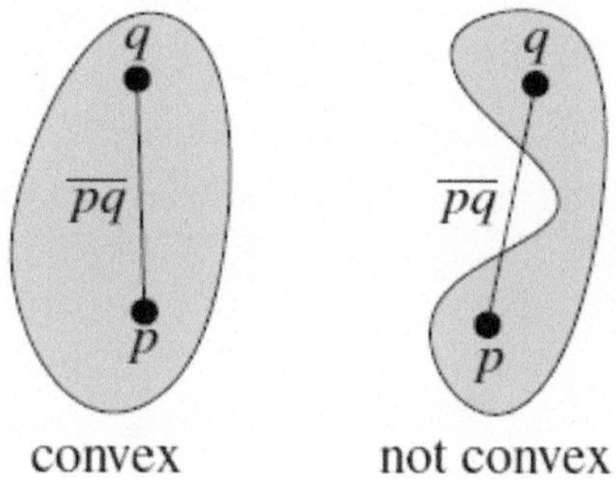

Figure 1.6. Difference between convex and non-convex points.

Source: https://www.researchgate.net/figure/a-Convex-Hull-b-Non-convex-Hull_fig2_267637542.

If a pair of points p,q∈ S the segment of line pq has been included inside it is said to be convex a subset of the plane. A convex hull CH(S) exists for a set S, which is the least convex set that contains the set S. To be more exact, it's the intersection of all convex sets that include the letter 'S' (Barber et al., 1993).

The subject of calculating the convex hull of a finite set 'P' of 'n' points in the plane would be investigated. A thought experiment may help us to visualize what the convex hull seems like. As though the points were nails protruding from the plane, obtain a stretchy band of rubber and wrap it on the nails before letting it go. This would crack around the nails and shorten them (Avis et al., 1992). The convex hull of 'P' is encompassed by the rubber band. As a consequence, the convex hull of a limited collection 'P' of points in the plane may be defined as the distinctive convex polygon its vertices have been points from 'P' and that comprises every point of P. We must verify that it has been properly described, which has been the distinct polygon and that the description is identical to the one provided previously, and we'll skip that for now (Sugihara, 2000).

What is the procedure for calculating the convex hull? We need to ask another question before we may answer this one: what is the meaning of the calculation of the convex hull? P's convex hull is a convex polygon, as we've shown. Listing the vertices of a polygon in the sequence of clockwise, starting with a random one, is a logical approach to express it. The issue we wish to resolve is this: given a collection of points in the plane P = {p_1, p_2, ..., p_n}, calculate a list of points from 'P' which have been the vertices of CH(P), ordered in the sequence of clockwise (Mucke, 2009).

Input is equal to the set of points:

p1, p2, p3, p4, p5, p6, p7, p8, p9

Output is equal to the depiction of the convex hull (Figure 1.7):

p4, p5, p8, p2, p9

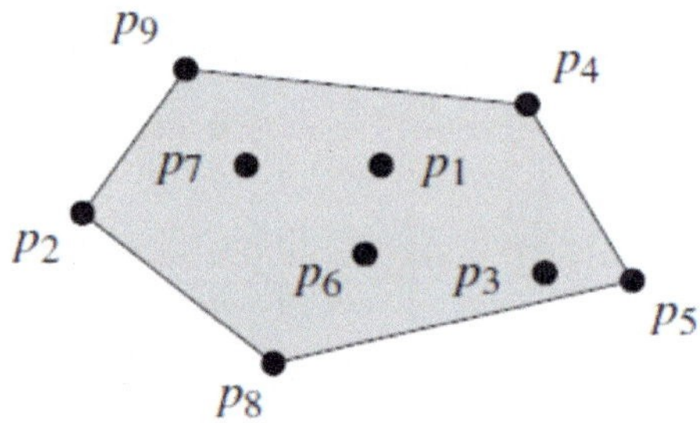

Figure 1.7. Computing a convex hull.

Source: https://www.chegg.com/homework-help/line-convex-hull-problem-given-set-q-n-points-one-point-time-chapter-33.3-problem-5e-solution-9780262032933-exc.

When we wish to create an algorithm to calculate the convex hull, the initial definition of convex hulls isn't very helpful. It is concerned with the crossover of all convex sets comprising P, of which there is an unlimited number. It is more important to notice that CH(P) is a convex polygon. Suppose that the borders of CH(P). Both of these edge endpoints 'p' and 'q' have been the 'P' points, and when we express the line by 'p' and 'q' so that CH(P) lies to the right side, after that, every of P's points should be to the right side of this line. The converse is also factual: if all of the points P\{p,q} are to the right side of the straight line via 'p' and 'q,' then 'pq' is a CH(P) edge (Chan et al., 2010).

We may construct an algorithm now that we have a better understanding of the issue's shape. We'll explain it via pseudo-code.

- **Algorithm SLOW CONVEX HULL(P):**

Input: The plane has a set of points P.

Output: In a clockwise sequence, a list 'L' contains the vertices of CH(P) (Fabri et al., 2000).

- E ← 0/;
- Whereas the arranged pairs (p,q) ∈ P×P having 'p' is not equal to q;
- Do valid ← true;
- Whereas all points r ∈ P are not equal to 'p' or 'q;'
- Do if 'r' lies to the left side of the directed line from 'p' to 'q;'
- After that valid ← false;
- if valid then Add up the directed edge pq→ to E;
- From the set 'E' of edges make a list 'L' of vertices of CH(P), arranged in the sequence of clockwise.

There are two stages in the algorithm that is possibly not evident to the reader (Bhagavathi et al., 1996). Line five: There is the first problem is addressed: how can we determine if a point is to the left side or right side of a directed line? The fundamental operations are needed by the vast majority of geometric algorithms. We suppose that certain operations are accessible at any point in time.

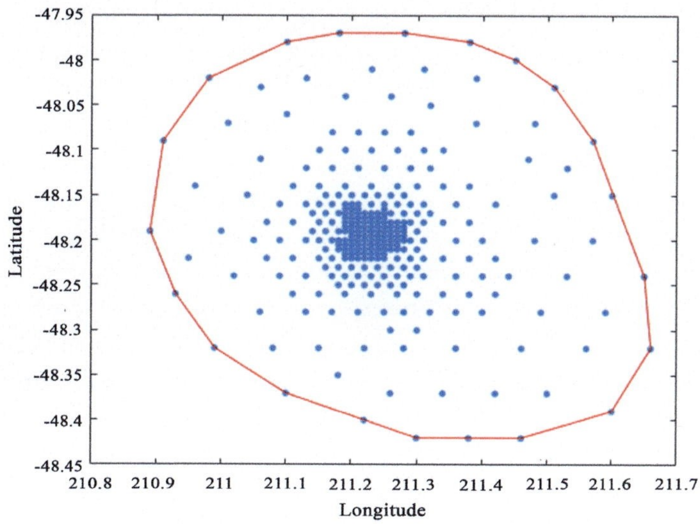

Figure 1.8. Computation of convex hull using MATLAB.

Source: https://www.mathworks.com/help/matlab/math/computing-the-convex-hull.html.

They may be completed in constant time, which means that the actual implementation would have no significant impact on the asymptotic running time in terms of size (Brun et al., 2012). To argue that these fundamental procedures are minor or trivial would be an understatement. They have been difficult to implement properly, and the time it takes to implement them would have an impact on the actual running time of the algorithms. Furthermore, software libraries comprising rudimentary functions like these are now widely accessible on the market. Following this conclusion, we can safely assume that we have a function accessible to us that will complete the test for us in constant time; hence, we do not need to worry about line 5 (Figure 1.8) (Tang et al., 2012).

The final phase of the method, which requires some explanation, is the other step that requires some explanation. It is determined in the loop of lines 2 to 7 how many convex hull edges there are in the set E. The following is how we may generate the list 'L' from the letter "E." Because the edges in 'E' are directed, we may refer to the starting point and endpoint of an edge as well as its origin and destination. Since the edges are directed after their origination while the vertices have been arranged in the sequence of clockwise, the edges are directed after their origin. Eliminate a random edge $\vec{e_1}$ from 'E' at this point (Hossain et al., 2013; Forrest, 2012). Insert the starting point of$\vec{e_1}$ as the initial point in 'L' and the end as the final point in 'L' to complete the equation. Locate the edge $\vec{e_2}$ in 'E' its initial point is similar to the endpoint of $\vec{e_2}$, delete it from E, and attach the endpoint of$\vec{e_1}$ to the end of the edge in L. After that, locate the edge $\vec{e_3}$ its initial point is the endpoint of $\vec{e_2}$, delete it from E, and append the endpoint of that edge to the end of L. We will carry on until there has been just a single edge remaining in 'E' (the last edge). Afterwards, we have been finished; the endpoint of the remaining edge has been the initial point of $\vec{e_1}$, which has been the initial point on line L. It takes $O(n^2)$ time to complete a basic implementation of this technique. Even if the time necessary for the remainder of the algorithm could be reduced to O(nlogn), the time necessary for the remaining of the algorithm would still dominate the overall time of operation (Chan et al., 2009).

SLOW CONVEX HULL's temporal complexity is simple to analyze. We examine $n^2 - n$ point pairings. We examine the n−2 additional points for every pair to determine if they all lie on the correct side. In all, this would need $O(n^3)$ time. Because the last step needs $O(n^2)$ time, the overall execution time is $O(n^3)$. Except for the smallest input sets, a method with a cubic operation time has been very sluggish and useful. The difficulty is that we didn't apply any creative algorithmic design methodologies; instead, we used brute force to transfer the geometric understanding into an algorithm. However, while we try to improve the algorithm, we should make a few observations about it (Czumaj et al., 2000).

When developing the condition for which a pair of 'p' and 'q' constitutes an edge of CH(P), we were a little hasty. There is a point 'r' that has not to lie to the right-side or left-side of the line connecting 'p' and 'q;' it may also be on this line. So, what are our options? This is referred to as a degenerate case, or simply degeneracy.

While we initially consider a problem, we prefer to overlook such circumstances so that we are not confused when attempting to determine the geometric aspects of the problem. In actuality, though, these circumstances do arise. When we construct the points by the click upon a screen with the help of a mouse, for example, every point would have tiny integer coordinates, and we would almost certainly end up with three points on a line (Chan et al., 2010).

We should reformate the criteria to make the algorithm valid in the presence of degeneracy: when each points 'r' does not lie precisely to the right-side of the directed line between 'p' and 'q,' or upon the open line-segment of 'p' and 'q,' a directed edge pq has been an edge of CH(P). (In P, we suppose there have been no points of coinciding.) As a result, we'll need to change line 5 of the algorithm more with this difficult test (Reif et al., 1989).

The following is a reworded version of the criterion above that may be used to ensure that the method is valid in the existence of degeneracy: When a directed edge pq→ has been an edge of CH(P), it means that all further points r $\in$ P are located either purely to the right side of the line which is directed through 'p' and 'q' or upon the open segment of line pq.

A directed edge pq is not an edge of CH(P) unless and until all other points are located on the open line segment pq. There have been no corresponding points in P, therefore we infer that there are none. As a result, line 5 of the algorithm must be replaced with this more sophisticated test (Figure 1.9) (Hershberger et al., 2003).

We are oblivious to another crucial factor that might have an impact on the validity of the outcome of our method, which we will discuss below. We made the implicit assumption that we would be able to determine whether a point is to the right side or to the left side of a particular line in a certain precise way. If the points have been supplied in floating-point coordinates and the calculations are performed utilizing floating-point arithmetic, there would be rounding mistakes that can skew the results of the evaluations (Samuel et al., 2000).

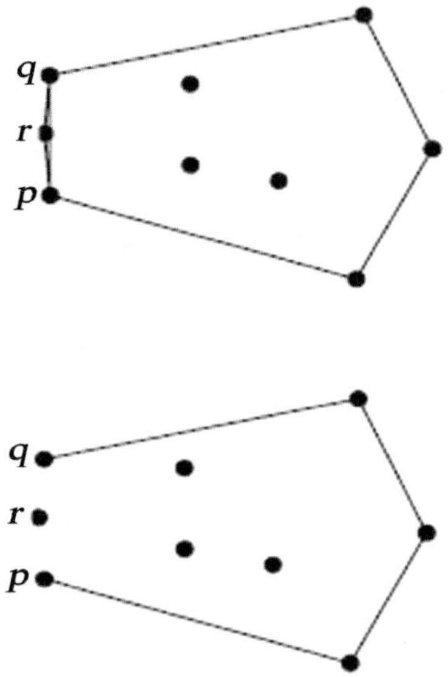

Figure 1.9. Illustration of pairing of points.

Source: https://www.amazon.com/Computational-Geometry-Applications-Mark-Berg/dp/3540779736.

Consider the following scenario: There have been three points 'p,' 'q,' and 'r' which have been almost on the same line, and other such points are located far to the right of such point. Our method examines the pairings (p,q), (r,q), and (p,q) (p,r). As a result of the proximity of these locations, the rounding errors can guide us to conclude that 'r' is located to the right side of the line connecting 'p' and q, that "p" is located to the right of the line connecting 'r' to 'q,' and that 'q' is located to the right of the line connecting 'p' and 'r.' Although it is not possible from a geometrical standpoint, the floating-point arithmetic has not recognized this fact (Matoušek et al., 1998). In this situation, the algorithm would accept all three edges as valid input parameters. The worst-case scenario is when all three tests return the opposite result, wherein the case algorithm eliminates all three edges, resulting in a gap in the border of the convex hull's perimeter. Thus, whenever we attempt to generate the arranged list of convex hull vertices in the final stage of our method, we run into a significant difficulty that we cannot solve. This phase imagines that there has been a single edge beginning in each convex hull vertex and that there has been precisely a single edge terminating in each convex hull vertex. Because of rounding problems, there may be two or no edges opening in vertex 'p' at any point in the graph. This has the potential to cause the software that implements our basic algorithm to fail because the final step was not planned to cope with this incompatible information in the first place (Sen, 1997).

Even though we have demonstrated that the method is valid and may control every particular instance, this has not been robust: even little flaws in the calculations may cause it to be unsuccessful in entirely unpredicted manners as we have demonstrated. There is a difficulty in that we have demonstrated correctness under the assumption that we may calculate with accurate real numbers (Figure 1.10).

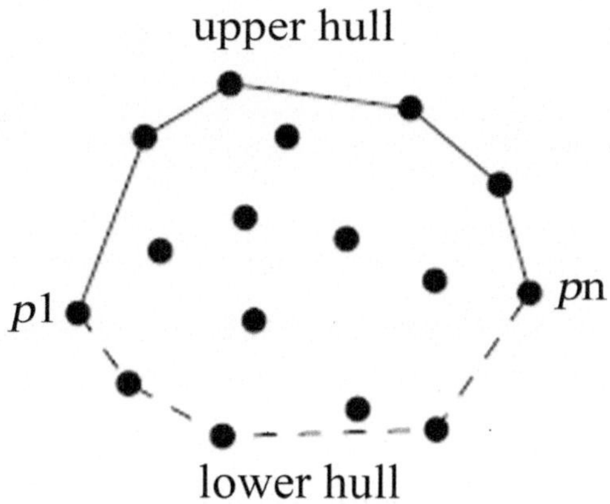

Figure 1.10. Illustration of upper and lower hulls.

Source: https://www.amazon.com/Computational-Geometry-Applications-Mark-Berg/dp/3540779736.

The initial geometric algorithm is created. It calculates the convex hull of the set of planar points. That is, although, rather sluggish; its time of execution is $O(n^3)$, it handles deteriorate instances awkwardly, and it is not resilient. We should make an effort to improve (Fukuda et al., 2001).

We'll use a typical algorithmic design method to do this: we'll create an algorithm. It implies that we'll insert the points in 'P' once at a time, revising our answer as we go. In adding up the points from left-side to right-side and give this incremental technique a geometric flavor. As a result, we initially arrange the points via x-coordinate, resulting in an arranged order p, which we afterwards add up in that sequence (Sugihara et al., 2000). Since we're functioning from left-side to right-side, it'd be helpful if the convex hull vertices, which appear along the boundary, were likewise arranged from left to right. This, however, is not the case. As a result, we only calculate the vertices of the convex hull that have been on the higher hull, which has been the component of the convex hull that runs from the left point p_1 to the right point p_n in the sequence of clockwise. The convex hull edges that surround the convex hull from above are contained in the upper hull. We calculate the rest component of the convex hull, the inferior hull, in a 2nd scan that runs from right-side to left-side (Figure 1.11) (Brown, 1979).

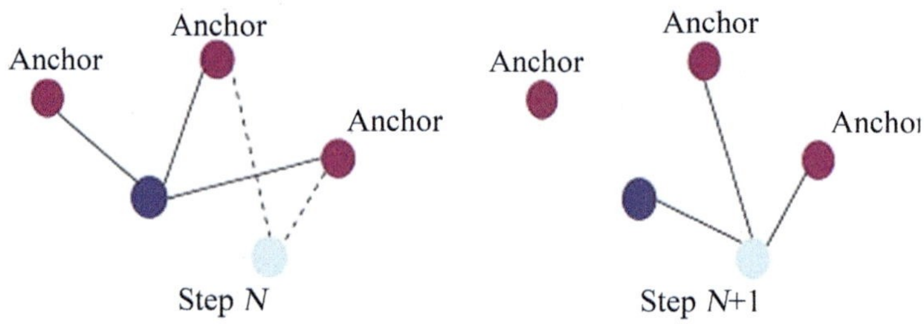

Figure 1.11. Incremental algorithm.

Source: https://www.sciencedirect.com/topics/engineering/incremental-algorithm.

The modification of the superior hull upon inserting a point p_i is the first stage in the incremental process. In other terms, we should calculate the upper hull of p_1, …, p_i –1 given the upper hull of the points p_1, …, p_i. p_1, …, p_i. It may be accomplished in the following manner. We make a turn at every vertex when walking around the polygon's border in a clockwise direction. This may be both a right and a left turn for an arbitrary polygon, however for a convex polygon; every turn must be a right turn. This recommends that you approach pi addition in the following manner (Boissonnat et al., 2007). Suppose that L_{upper} is a list that keeps track of the upper vertices from left to right. L_{upper} is initially appended with p_i. P_i has been the right-side point; therefore, it should be on the top hull. After that, we look to see if L_{upper} has the final three points to form a right turn. If that's the case, we don't need to do anything further; Lupper includes the components of the upper hull of p_1, …, p_i, and we may go on to the subsequent point, p_i+_1. However, if the final three points form a left turn, the center one must be removed from the top hull. We're not done yet in this situation: it's possible that the novel final three points still don't form a right turn, wherein the case we must eliminate the middle one once again (Chan et al., 2001). We keep going in this direction until the final three points turn right or there have been two points remaining.

The algorithm is now shown in pseudo-code. Both the lower and upper hulls are computed using the pseudo-code. The final is accomplished via processing the points from right-side to left-side, similar to how the upper hull is computed (Barber, 1992).

- **Algorithm of CONVEX HULL(P):**

Input: The plane has a set 'P' of points.

Output: The components of CH(P) are listed in the sequence of clockwise in this list (Boissonnat et al., 1992).

- Arrange the points via x-coordinate, resultant in the order p_1, …, p_n;
- Insert the points p_2 and p_1 in the L_{upper} list, having p_1 as the initial point;
- Whereas I ← 3 towards n;
- Add up the p_i to L_{upper} list;
- Where L_{upper} contains greater than two points and the final three points in the L_{upper} list don't form a right turn;
- Eliminate the center of the final three points from the list L_{upper};
- Insert the points p_n −1 and p_n in L_{lower} list, having the p_n as the initial point;
- Whereas i← n−2 downwards 1;
- Add up the p_i to the list L_{lower};
- Whereas the list L_{lower} contains greater than two points and the final three points in the list L_{lower} and don't form a right turn;
- Eliminate the center of the final three points from the list L_{lower};
- Eliminate the initial and the final point from the list L_{lower} to keep away from repetition of the points whereas the lower and upper hull get together;
- Add up the list L_{lower} to the list L_{upper}, and call the resultant list L;
- Return to list L.

Whenever we examine the above algorithm more closely, we can see that it is incorrect. We presumed that no 2 locations have the same x-coordinate without saying it. The ordering on the x-coordinate is not clearly defined whether this assumption is false. However, this does not appear to be a severe issue. All we have to do now is correctly generalize the ordering: instead of utilizing the x-coordinate of the points to create the sequence, we'll utilize the sequence of lexicographic. This indicates that we arrange via x-coordinate first, and then by y-coordinate if the x-coordinates are the same (Imai and Iri, 1986). We also overlooked the fact that the three places for which we should evaluate if they execute a right or left turn are on a straight line. Because the center point must not be on the convex hull in this situation, points on the same line should be handled as though they are turning left. To put it another way, we must utilize a test that returns true if the three points turn right and false if they don't. (Note that this is a simple test than the one needed by the prior technique when collinear points had been present) (Klimenko et al., 2021).

In light of such modifications, the convex hull is properly computed by the algorithm. The upper hull has been described as that portion of the convex hull operating from the most lexico-graphically least vertex to the most lexico-graphically major vertex, and the remainder of the convex hull has been computed by the 2nd scan, which is described as that portion of the hull remaining after computing the upper hull (Lubiw et al., 2020).

What does our method do in the case of floating-point arithmetic rounding errors? When these mistakes happen, a point may be eliminated from the convex hull when it must not be, or a point within the true convex hull may not be eliminated at all. However, the algorithm's structural integrity is unaffected: this would calculate a polygonal closed chain. Even so, the result is the points list that we may read as a clockwise list of a polygon's components, as well as any three successive points that form a right turn or almost form a right turn due to rounding mistakes. Furthermore, no point in 'P' may be too distant from the calculated hull (Lee et al., 1995). The one issue that still exists is that a quick left turn might be misinterpreted as a right turn when three locations are extremely close together. The final polygon may have a dent as a result of this. One approach around this is to ensure that points in the input that have been extremely close together are treated as a similar point, for instance via rounding. As a consequence, while the outcome does not have to be accurate, we could not expect a precise output if we employ exact arithmetic, which makes sense. This is sufficient for several applications. Still, it's an excellent idea to be cautious when running the fundamental test to prevent making mistakes.

We'll wrap things off with the given below theorem (Wu et al., 2008):

- **Theorem 1.1:** In O(nlogn) time, the convex hull of a set of 'n' points in the plane may be calculated.

Proof: We'll show that the upper hull computation is valid; similar reasons may be used to show that the lower hull computation is right. The evidence is based upon the number of sites that have been handled (Bandyopadhyay et al., 2007). The point's p_1 and p_2, which comprises the upper hull of p_1, and p_2, are in the list L_{upper} before the for-loop starts. Assume the addition of p_i to L_{upper} which includes the upper hull vertices of $\{p_1, \ldots, p_i - 1\}$. We understand that the points in L_{upper} make a chain that only takes right turns once the while-loop is executed and because of the induction theory. Furthermore, the chain begins with the lexicographically smallest point, $\{p_1, \ldots, p_i\}$, and concludes with the lexicographically biggest point, p_i. When we may prove that all $\{p_1, \ldots, p_i\}$ points that aren't in L_{upper} have been beneath the chain after that L_{upper} includes the proper points. We understand that there has been no point over the chain which existed before p_i was introduced through induction (Assarf et al., 2017). Because the old chain is lower than the novel chain, the only way a point may be higher than the novel chain, is in the vertical slab between p_i-1 and p_i. However, because a point will lie between p_i and p_i-1 in the lexicographical sequence, this is not feasible. (If p_i-1 and p_i, or other points, with the similar x-coordinate, you need to show that similar reasoning holds) (Salesin et al., 1989).

To demonstrate the time-bound, we notice that lexicographically organizing the points may be completed in the time of O(nlogn). Now examine the upper hull's calculation. The for-loop is run a certain time. The last question is how frequently the while-loop within it has been run (Fabri et al., 2000). The while-loop is run at least once per for-loop implementation. A point is deducted from the existing hull for each extra execution. Because every point may only be removed once during the top hull's creation, the overall number of additional executions for all for-loops is limited to n. The calculation of the bottom hull requires O(n) time as well. The overall time needed to calculate the convex hull is owing to the sorting phase O(nlogn) (Halpering, 2002). Finally, the ultimate convex hull technique is straightforward to define and straightforward to execute. There are just two steps required: lexicographic sorting and a check to see if three successive points are in the appropriate direction. At first glance, it was difficult to imagine that this simple and effective solution could exist given the initial statement of the issue (Toussaint, 1994).

1.4. DEGENERACIES AND ROBUSTNESS

The construction of a geometric algorithm frequently consists of three phases. We strive to disregard everything that would obstruct our grasp of the geometric principles we're working on within the initial phase. Vertical line segments, like collinear points, maybe a pain in the neck. It's frequently easier to overlook such degenerate circumstances when 1st trying to develop or comprehend an algorithm (Figure 1.12) (Berg et al., 1997; Megerian et al., 2005).

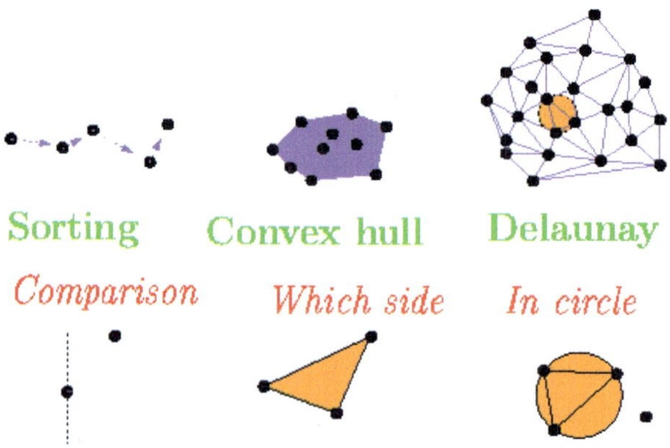

Figure 1.12. Illustration of robustness and degeneracies.

Source: https://www-sop.inria.fr/prisme/fiches/Robustesse/index.html.en.

We must change the algorithm created in the 1st step to be correct in the existence of degenerate situations in the 2nd phase. Beginners usually achieve this by overloading their algorithms with case distinctions. There seems to be a better option in several cases. It is typically possible to merge specific circumstances with the generalized form by revisiting the issue's geometry (Biasotti et al., 2008). To cope with points having equal x-coordinates, we merely had to utilize the lexicographical sequence rather than the order on x-coordinate in the convex hull algorithm. We've tried to use this integrated method to cope with unusual circumstances in most of the algorithms. Even said, it's simpler not to consider these scenarios when reading the book for the 1st time. Only when you've figured out how is the working of algorithm, generally, you must consider the degeneracy (Schirra, 1998).

If you go through the literature on calculational geometry, you'll notice that several writers disregard particular instances, frequently by establishing specific input hypotheses we might have omitted special circumstances in the convex hull issue via merely declaring that we suppose the input has been that having no 3 points are on the same line and also having no 2 points the similar x-coordinate. These assumptions are frequently justified from a theoretical standpoint: the purpose is to determine the computational difficulty of a problem, and while working out the specifics might be tiresome, degenerate circumstances may be controlled without raising the algorithm's asymptotic complication. Special situations, on the other hand, add to the implementation's complexity (Schorn, 1991).

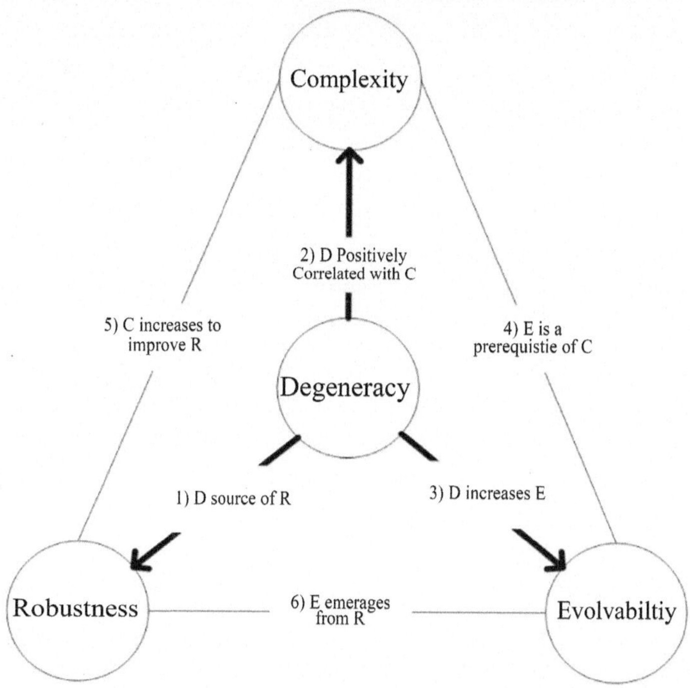

Figure 1.13. High level illustration of the relationships between degeneracy, complexity, robustness, and evolvability.

Source: https://www.researchgate.net/figure/high-level-illustration-of-the-relationships-between-degeneracy-complexity-robustness_fig1_215617336.

Most computational geometry researchers nowadays are conscious that their basic position hypotheses are not met in practice, and that treating exceptional instances as a whole is usually the best method to deal with them. Furthermore, there are generic approaches known as representative perturbation plans that enable one to disregard particular circumstances during the execution and design of an algorithm while still producing a proper algorithm in the existence of degeneracy (Figure 1.13) (Torr et al., 1998).

The actual implementation is the 3rd phase. Now it's time to consider the fundamental operations, such as determining if a point is to the left-side, right-side, or on a directed line. If you're lucky, you'll find a geometric software library that has the operations you want; otherwise, you'll have to code them manually (Schirra, 1996).

An additional difficulty that occurs during the execution stage is that the premise of executing accurate arithmetic having real numbers fails, and the ramifications must be understood. Whenever it comes to geometric algorithms, robustness issues are a common source of irritation. It's not simple to solve problems of robustness. One option is to utilize software that performs precise arithmetic (utilizing rationales, integers, or even algebraic numbers, dependent upon the sort of issue), however, this would be time-consuming (Kettner et al., 2008; Yap, 1997). Instead, the algorithm may be tweaked to detect errors and take necessary action to prevent the application from crashing.

It is vital to specify the exact qualities of the result in this scenario since the algorithm's output is not guaranteed to be right. This is what we accomplished when we constructed the convex hull method in the preceding part: the outcome may not be a convex polygon, however, we understand the output structure is right, and the resultant polygon is extremely near to the convex hull. Ultimately, depending on the input, the accuracy in the number description necessary to resolve the issue properly may be predicted (Marshall et al., 2001).

Depending on the application, which strategy is preferable? Exact arithmetic is chosen if speed is not a problem. In other circumstances, the accuracy of the algorithm's outcome isn't as significant. While showing the convex hull of a group of points, for instance, it is unlikely that the polygon deviates somewhat from the genuine convex hull would be observed. We may utilize a cautious implementation depending upon floating-point arithmetic in this scenario (Mei et al., 2014).

1.5. APPLICATION DOMAINS

Each geometric topic, technique, or data structure covered in the section has a motivating example application. The majority of the applications come from the fields of robotics, graphics of computers, GIS, and CAM/CAD. We offer a quick summary of the regions and some of the geometric difficulties that emerge in them for those who are unfamiliar with them (Fabri et al., 2009).

1.5.1. Computer Graphics

The field of computer graphics deals with making pictures of modeled situations for presentation on a printer, the screen of a computer, or further output devices. The settings range from simple two-dimensional sketches of polygons, lines, and other rudimentary things to realistic-looking three-dimensional scenes with textures, light resources, and other details. Over a million patches of curved surface or polygons may readily be found in the latter sort of scenario (Tamassia, 1996).

Geometric algorithms are significant in computer graphics as sceneries are made up of geometric objects (Figure 1.14).

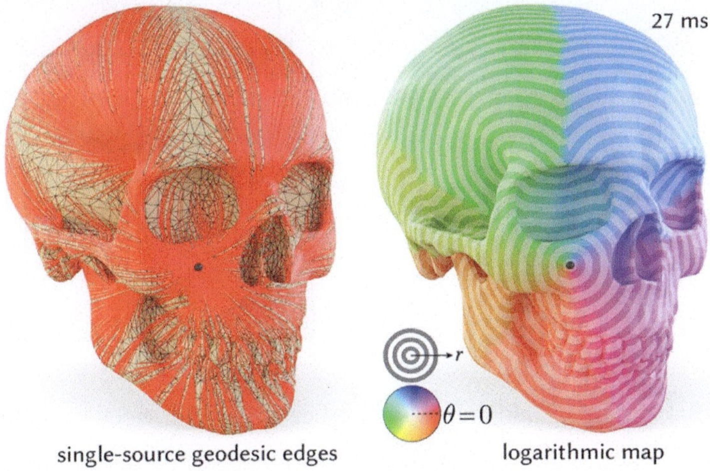

Figure 1.14. Application of computational geometry in computer graphics.

Source: http://graphics.cs.cmu.edu/.

The intersection of specific primitives, finding the primitive pointed to with the mouse, or identifying the subset of primitives that fall within a given region are all common issues in two-dimensional graphics (De Berg et al., 2008).

Geometric questions grow more complicated while coping with three-dimensional difficulties. Concealed surface elimination is a critical step in showing a three-dimensional picture: determining which elements of a scene are accessible from a certain viewpoint or, in other terms, discarding the bits that are hidden behind other objects (Barequet et al., 1997).

We must consider light while creating realistic-looking sceneries. This leads to a slew of additional issues, like shadow calculation. As a result, realistic picture synthesis necessitates the use of advanced display methods such as radiosity and ray tracing. It's critical to detect collisions among items while working with moving objects or in virtual reality applications. All of such scenarios have geometric issues (Figure 1.15) (De Berg et al., 2000).

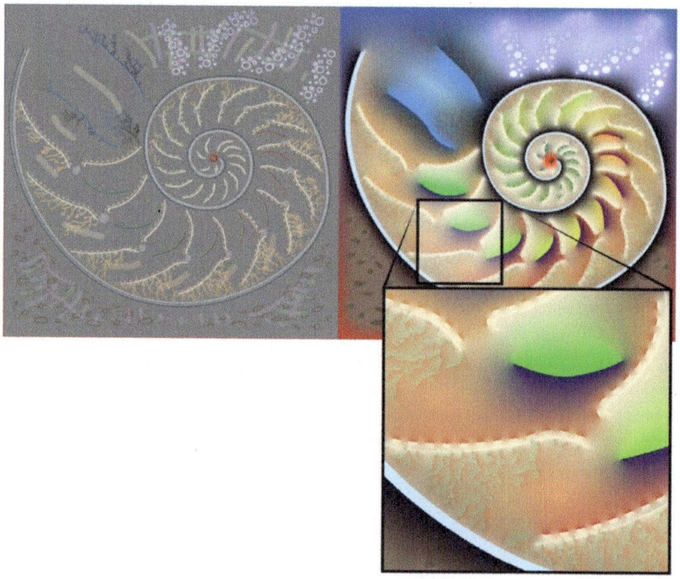

Figure 1.15. Computational geometry application in enhanced computer graphics.

Source: https://www.cs.cmu.edu/~kmcrane/Projects/MonteCarloGeometryProcessing/index.html.

1.5.2. Robotics

Robotics is the study of robot design and application. Because robots have been geometric things that work in three-dimensional space in the actual world, geometric issues emerge frequently. We discussed the motion planning issue earlier, in which a robot must find a way across an obstacle-filled environment. Motion planning is one facet of the larger task planning issue. Giving a robot the higher-level job of "vacuuming the room" and allowing the robot to find out the best approach to do the task will be ideal. This entails planning motions, as well as the sequence wherein subtasks, should be completed (Mouratidis, 2018).

Other geometric issues arise when designing robots and the work cells in which they must function. The majority of industrial robots are fixed-base robot arms. The parts that the robot arm works on must be given in this way that the robot may grab them effortlessly (Wan et al., 2005). Certain parts can need to be immobilized for the robot to work on them. Before the robot may operate on them, they may require to be rotated to a recognized orientation. These are all geometric issues, with a kinematic element in certain cases. A few of the algorithms can be used to solve similar situations. For instance, the smallest enclosing disc issue may be utilized to determine where robot arms should be placed (Figure 1.16) (Mitchell, 1996).

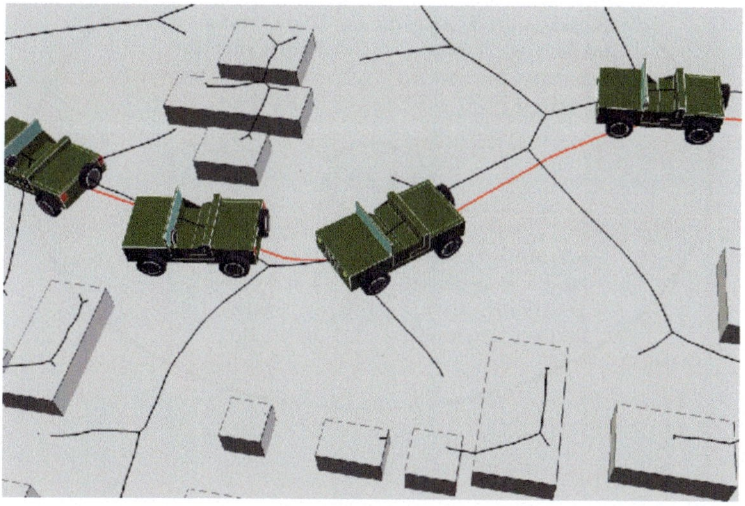

Figure 1.16. Illustration of computational geometry in robotics.

Source: https://www.semanticscholar.org/paper/USING-METHODS-OF-COMPUTATIONAL-GEOMETRY-IN-ROBOTICS-Svec/996fbb24c11a3f4e0456f23dad7156a6846ac674.

1.5.3. Geographic Information Systems (GIS)

A GIS holds geographical data such as country shapes, mountain heights, river courses, plant types at various places, population density, and rainfall.

Cities, highways, trains, power lines, and gas pipelines are examples of human-made constructions that may be stored. A GIS may be utilized to extract data about particular locations, as well as information about the relationships between various kinds of information (Mehlhorn et al., 1997). A biologist, for instance, could wish to link average rainfall to the presence of specific plants, and a civil engineer might require querying GIS to see whether there have been gas pipes beneath a lot where the work of excavation is to be done. Geometric difficulties abound because most geographic data is concerned with the qualities of points and areas on the earth's surface. Furthermore, because the volume of data is so huge, effective algorithms are required (Figure 1.17) (Mitchell, 1996).

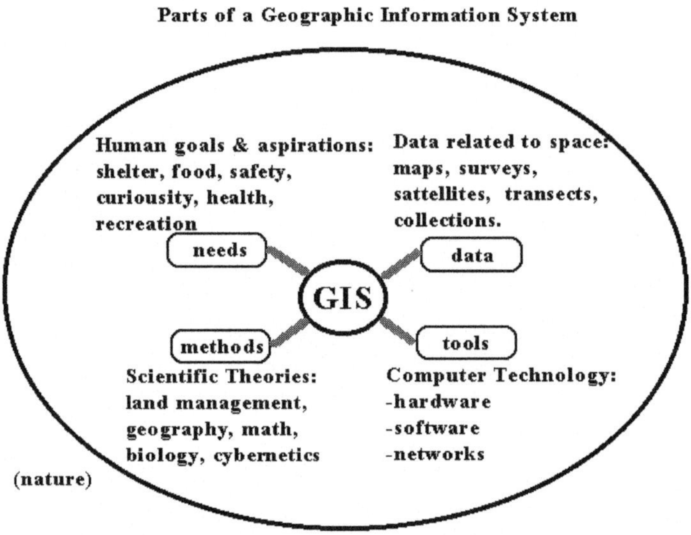

Figure 1.17. Illustration of GIS systems.

Source: https://www.conservationgis.org/gishistory/gishistry2.html.

The 1st concern is the storage of geographic data. Assume we would like to create an automobile navigation system that displays the driver where she would be at all times. This necessitates the storage of a massive map of roads as well as other data. We must be capable of rapidly detecting the location of the automobile on the map and picking a smaller area of the map for presentation on the onboard computer at all times. These procedures necessitate the use of efficient data structures (Sethian, 1999).

In some hilly regions, height information is frequently only accessible at a few sample places. For other positions, we'll have to interpolate between close sample points to get the heights. But how do we decide which specimen points to use? Among the most significant processes in a Geographic Information System is the merging of multiple kinds of information (De Floriani et al., 2000). For instance, we could wish to see which residences are in a forest, identify all bridges via looking at whereas highways cross rivers or find a decent location for a novel golf course via looking for a little hilly, relatively inexpensive land not away from a village. Different sorts of data are frequently stored in various maps in a Geographic Information System. We must overlay multiple maps to merge the data (Saračević et al., 2013).

Ultimately, we return to the identical instance from earlier in this chapter: the location of the nearby public mobile phone (hospital, or other facilities). This necessitates the creation of the diagram of Voronoi (Bernroider, 1978).

1.5.4. CAD/CAM

CAD is the process of creating goods using a computer. Printed circuit boards, machine components, and furniture, as well as full structures, are all possible goods. Because the final output is always a geometric object, all kinds of geometric issues are to be expected. CAD software must cope with object unions and crossovers, as well as decomposing objects and object boundaries into smaller forms and viewing the planned items (Figure 1.18) (Meguerdichian et al., 2001).

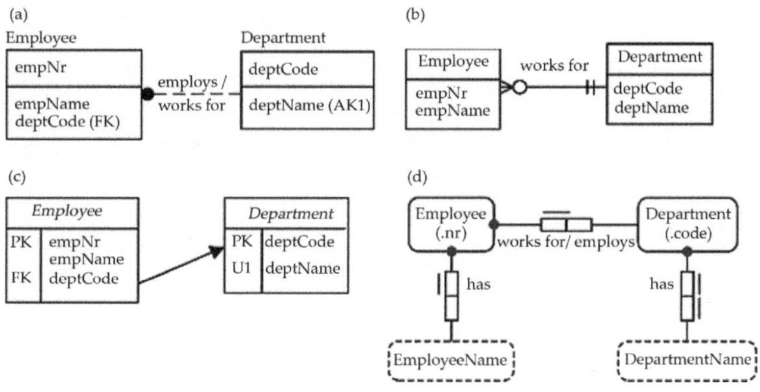

Figure 1.18. Sample examples of CAD/CAM applications (4 cases, a–d).

Source: https://www.sciencedirect.com/topics/computer-science/computer-aided-manufacturing.

Various tests are required to determine whether a design fits the standards. Often, a simulation will serve instead of building a prototype for such testing.

It is necessary to build a thing once it is developed and tested. Computer-aided manufacturing (CAM) software may help with this (Šeda et al., 2020; Chen et al., 2015).

1.5.5. CAM Issues

A newer trend is built for assembly, in which assembly considerations are made early in the design process. A Computer-Aided Design system that supports and will enable designers to investigate their viability of designs, answering issues such as: may the product be constructed quickly utilizing a specific manufacturing procedure? Answering several such problems necessitates the use of geometric algorithms (Cooper, 2013; Schurmann, 2009).

1.5.6. Domains for Other Applications

Geometric issues may occur in a variety of application disciplines, and geometric methods and data structures may be employed to address them.

Molecules, for instance, are commonly depicted in molecular modeling as the sets of interconnecting balls in space, single balls for every atom. Calculating the collection of the atom balls to generate the surface of a molecule or calculating wherein two molecules may contact one other are examples of common problems (Figure 1.19) (De Berg, 1997).

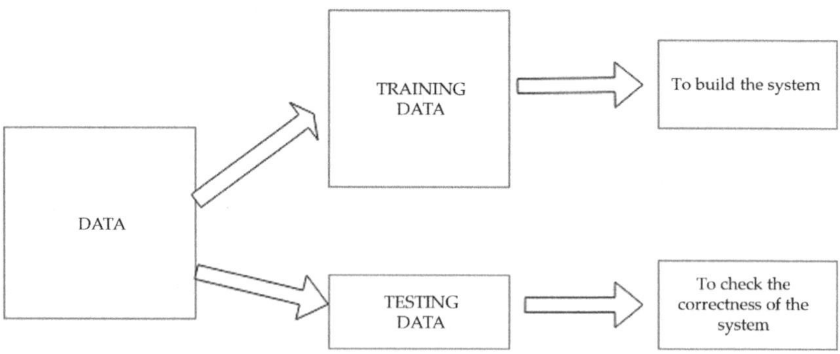

Figure 1.19. Pattern recognition flowchart.

Source: https://www.geeksforgeeks.org/pattern-recognition-introduction/.

The recognition of the pattern is another field. Assume a system that recognizes characters using optical character recognition. A system like this scans a piece of paper with text on it to recognize the characters. Matching a character's picture against a set of stored characters to discover one that finest suits it is a fundamental step.

Caffeine that leads to a geometric issue and identify how closely two geometric objects are like one other (Figure 1.20) (Ironi and Tentoni, 2012).

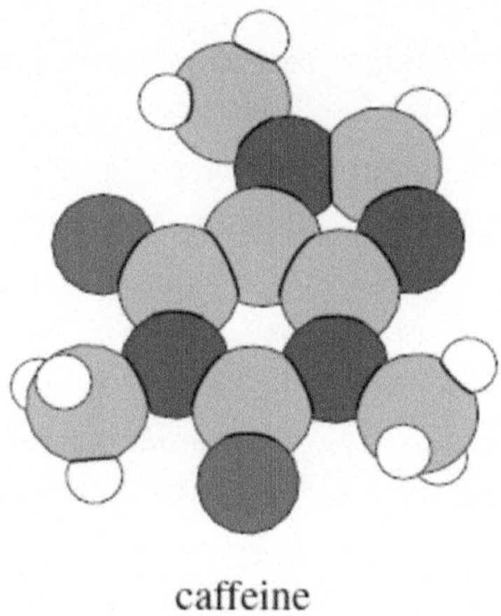

Figure 1.20. Structural representation of caffeine.

Source: https://link.springer.com/book/10.1007/978-3-540-77974-2.

Even domains that do not appear to be geometric at 1st glance may advantage from geometric algorithms since it is sometimes probable to describe a non-geometric issue in a geometrically oriented manner. As an example, we would show how the records in the data may be understood like points in a space of high-dimensional, and we would provide a data structure geometry that would allow specific queries NOTES AND COMMENTS on the records to be answered quickly (Poulos et al., 2012).

We believe that the set of geometric issues presented above has demonstrated that computational geometry has applications in a wide range of disciplines of computer science and engineering. The structures, algorithms, data, and strategies would provide you with the knowledge and skills necessary to successfully tackle geometric issues of this nature (Buchberger, 1988).

REFERENCES

1. Assarf, B., Gawrilow, E., Herr, K., Joswig, M., Lorenz, B., Paffenholz, A., & Rehn, T., (2017). Computing convex hulls and counting integer points with polymake. Mathematical Programming Computation, 9(1), 1–38.
2. Avis, D., & Fukuda, K., (1992). A pivoting algorithm for convex hulls and vertex enumeration of arrangements and polyhedra. Discrete and Computational Geometry, 8(3), 295–313.
3. Bandyopadhyay, D., & Snoeyink, J., (2007). Almost-Delaunay simplices: Robust neighbor relations for imprecise 3D points using CGAL. Computational Geometry, 38(1, 2), 4–15.
4. Barber, C. B., (1992). Computational Geometry with Imprecise Data and Arithmetic (Vol. 2, No. 2, pp. 10–15). Princeton University.
5. Barber, C. B., Dobkin, D. P., & Huhdanpaa, H., (1993). The Quickhull Algorithm for Convex Hull (Vol. 20, No. 3, pp. 368–380). Technical Report GCG53, The Geometry Center, MN.
6. Barber, C. B., Dobkin, D. P., & Huhdanpaa, H., (1996). The quickhull algorithm for convex hulls. ACM Transactions on Mathematical Software (TOMS), 22(4), 469–483.
7. Barequet, G., Bridgeman, S. S., Duncan, C. A., Goodrich, M. T., & Tamassia, R., (1997). Classical computational geometry in GeomNet. In: Proceedings of the Thirteenth Annual Symposium on Computational Geometry (3rd edn., pp. 412–414).
8. Berg, M. D., Kreveld, M. V., Overmars, M., & Schwarzkopf, O., (1997). Computational geometry. In: Computational Geometry (Vol. 1, pp. 1–17). Springer, Berlin, Heidelberg.
9. Bernroider, G., (1978). The foundation of computational geometry: Theory and application of the point-lattice-concept within modern structure analysis. In: Geometrical Probability and Biological Structures: Buffon's 200th Anniversary (Vol. 1, pp. 153–170). Springer, Berlin, Heidelberg.
10. Bhagavathi, D., Gurla, H., Olariu, S., Schwing, J. L., & Zhang, J., (1996). Square meshes are not optimal for convex hull computation. IEEE Transactions on Parallel and Distributed Systems, 7(6), 545–554.
11. Biasotti, S., Giorgi, D., Spagnuolo, M., & Falcidieno, B., (2008). Reeb graphs for shape analysis and applications. Theoretical Computer Science, 392(1–3), 5–22.
12. Boissonnat, J. D., & Teillaud, M., (2007). Effective Computational Geometry for Curves and Surfaces (Vol. 2, No. 2, pp. 10–15). Heidelberg: Springer.

13. Boissonnat, J. D., Devillers, O., Schott, R., Teillaud, M., & Yvinec, M., (1992). Applications of random sampling to on-line algorithms in computational geometry. Discrete & Computational Geometry, 8(1), 51–71.
14. Brown, K. Q., (1979). Voronoi diagrams from convex hulls. Information Processing Letters, 9(5), 223–228.
15. Brun, C., Dufourd, J. F., & Magaud, N., (2012). Designing and proving correct a convex hull algorithm with hypermaps in Coq. Computational Geometry, 45(8), 436–457.
16. Buchberger, B., (1988). Applications of gröbner bases in non-linear computational geometry. In: Mathematical Aspects of Scientific Software (4th edn., pp. 59–87). Springer, New York, NY.
17. Chan, T. M., & Chen, E. Y., (2010). Optimal in-place and cache-oblivious algorithms for 3D convex hulls and 2D segment intersection. Computational Geometry, 43(8), 636–646.
18. Chan, T. M., & Pa ˇtraşcu, M., (2009). Transdichotomous results in computational geometry, I: Point location in sublogarithmic time. SIAM Journal on Computing, 39(2), 703–729.
19. Chan, T. M., & Patrascu, M., (2010). Transdichotomous Results in Computational Geometry, II: Offline Search (Vol. 3, No. 2, pp. 6–10). arXiv preprint arXiv:1010.1948.
20. Chan, T. M., (2001). Dynamic planar convex hull operations in near-logarithmic amortized time. Journal of the ACM (JACM), 4(3), 1–6.
21. Chen, D. Z., & Wang, H., (2015). Visibility and ray shooting queries in polygonal domains. Computational Geometry, 48(2), 31–41.
22. Cooper, H. M., (2013). Discrete Groups and Computational Geometry: A Thesis Presented in Partial Fulfilment of the Requirements for the Degree of Doctor of Philosophy in Mathematics at Massey University, Albany, New Zealand (Vol. 3, No. 2, pp. 3–7). Doctoral dissertation, Massey University.
23. Czumaj, A., Sohler, C., & Ziegler, M., (2000). Property testing in computational geometry. In: European Symposium on Algorithms (Vol. 2, pp. 155–166). Springer, Berlin, Heidelberg.
24. De Berg, M. T., Van, K. M., Overmars, M., & Schwarzkopf, O., (2000). Computational geometry: Algorithms and Applications (Vol. 2, No. 1, pp. 5–15). Springer Science & Business Media.
25. De Berg, M., (1997). Trends and developments in computational geometry. In: Computer Graphics Forum (Vol. 16, No. 1, pp. 3–30). Oxford, UK and Boston, USA: Blackwell Publishers Ltd.

26. De Berg, M., Cheong, O., Van, K. M., & Overmars, M., (2008). Computational geometry: Introduction. Computational Geometry: Algorithms and Applications, 5(4), 1–17.
27. De Berg, M., Cheong, O., Van, K. M., & Overmars, M., (2008). Computational geometry: Introduction. Computational Geometry: Algorithms and Applications, 1, 1–17.
28. De Floriani, L., Magillo, P., & Puppo, E., (2000). Applications of computational geometry to geographic information systems. Handbook of Computational Geometry, 7, 333–388.
29. Fabri, A., & Pion, S., (2009). CGAL: The computational geometry algorithms library. In: Proceedings of the 17th ACM SIGSPATIAL International Conference on Advances in Geographic Information Systems (4th edn., pp. 538–539).
30. Fabri, A., Giezeman, G. J., Kettner, L., Schirra, S., & Schönherr, S., (2000). On the design of CGAL a computational geometry algorithms library. Software: Practice and Experience, 30(11), 1167–1202.
31. Forrest, A. R., (2012). Computational Geometry Project School of Information Systems (Vol. 3, No. 2, p. 113). Fundamental Algorithms for Computer Graphics: NATO Advanced Study Institute directed by JE Bresenham, RA Earnshaw, MLV Petteway.
32. Fukuda, K., Liebling, T. M., & Lütolf, C., (2001). Extended convex hull. Computational Geometry, 20(1, 2), 13–23.
33. Ghosh, S., (2000). Electromagnetic field theory as a basis for the odd parity rule in computational geometry. International Journal of Engineering Education, 16(1), 68–72.
34. Goodrich, M. T., Tsay, J. J., Vengroff, D. E., & Vitter, J. S., (1993). External-memory computational geometry. In: Proceedings of 1993 IEEE 34th Annual Foundations of Computer Science (pp. 714–723). IEEE.
35. Gu, D. X., Luo, F., & Yau, S. T., (2010). Fundamentals of computational conformal geometry. Mathematics in Computer Science, 4(4), 389–429.
36. Halperin, D., (2002). Robust geometric computing in motion. The International Journal of Robotics Research, 21(3), 219–232.
37. Held, M., (1991). On the Computational Geometry of Pocket Machining (Vol. 500, pp. 5–10). Springer Science & Business Media.
38. Hershberger, J., & Suri, S., (2003). Convex hulls and related problems in data streams. In: Proc. of the ACM/DIMACS Workshop on Management and Processing of Data Streams (Vol. 136, pp. 2, 3).
39. Hossain, M. Z., & Amin, M. A., (2013). On Constructing Approximate Convex Hull, 3(2), 11–15.

40. Imai, H., & Iri, M., (1986). Computational-geometric methods for polygonal approximations of a curve. Computer Vision, Graphics, and Image Processing, 36(1), 31–41.
41. Ironi, L., & Tentoni, S., (2012). Interplay of spatial aggregation and computational geometry in extracting diagnostic features from cardiac activation data. Computer Methods and Programs in Biomedicine, 107(3), 456–467.
42. Kettner, L., Mehlhorn, K., Pion, S., Schirra, S., & Yap, C., (2008). Classroom examples of robustness problems in geometric computations. Computational Geometry, 40(1), 61–78.
43. Klimenko, G., Raichel, B., & Van, B. G., (2021). Sparse convex hull coverage. Computational Geometry, 98, 101–787.
44. Lee, Y. S., & Chang, T. C., (1995). Application of computational geometry in optimizing 2.5 D and 3D NC surface machining. Computers in Industry, 26(1), 41–59.
45. Lubiw, A., Maftuleac, D., & Owen, M., (2020). Shortest paths and convex hulls in 2D complexes with non-positive curvature. Computational Geometry, 89, 101626.
46. Marshall, D., Lukacs, G., & Martin, R., (2001). Robust segmentation of primitives from range data in the presence of geometric degeneracy. IEEE Transactions on Pattern Analysis and Machine Intelligence, 23(3), 304–314.
47. Matoušek, J., & Plecháč, P., (1998). On functional separately convex hulls. Discrete & Computational Geometry, 19(1), 105–130.
48. Megerian, S., Koushanfar, F., Potkonjak, M., & Srivastava, M. B., (2005). Worst and best-case coverage in sensor networks. IEEE Transactions on Mobile Computing, 4(1), 84–92.
49. Meguerdichian, S., Koushanfar, F., Potkonjak, M., & Srivastava, M. B., (2001). Coverage problems in wireless Ad-Hoc sensor networks. In: Proceedings IEEE INFOCOM 2001; Conference on Computer Communications; Twentieth Annual Joint Conference of the IEEE Computer and Communications Society (Cat. No. 01CH37213) (Vol. 3, pp. 1380–1387). IEEE.
50. Mehlhorn, K., Müller, M., Näher, S., Schirra, S., Seel, M., Uhrig, C., & Ziegler, J., (1997). A computational basis for higher-dimensional computational geometry and applications. In: Proceedings of the Thirteenth Annual Symposium on Computational Geometry (Vol. 4, No. 3, pp. 254–263).

51. Mei, G., Tipper, J. C., & Xu, N., (2014). Numerical robustness in geometric computation: An expository summary. Applied Mathematics & Information Sciences, 8(6), 2717.
52. Mitchell, J. S., (1996). On some applications of computational geometry in manufacturing and virtual environments. In: Workshop on Applied Computational Geometry (2nd edn., pp. 37–40). Springer, Berlin, Heidelberg.
53. Mitchell, J. S., (1996). On some applications of computational. In: Applied Computational Geometry. Towards Geometric Engineering: FCRC'96 Workshop, WACG'96 (Vol. 114, p. 37). Philadelphia, PA, Selected Papers. Springer Science & Business Media.
54. Mouratidis, K., (2018). Applying spatial database techniques to other domains: A case study on top-k and computational geometric operators. In: Proceedings of the Fifth International ACM SIGMOD Workshop on Managing and Mining Enriched Geo-Spatial Data (Vol. 2, No. 3, pp. 25–26).
55. Mucke, E., (2009). Computing prescriptions: Quickhull: Computing convex hulls quickly. Computing in Science & Engineering, 11(5), 54–57.
56. Noltemeier, H., (1988). Computational geometry and its applications CG'88, international workshop on computational geometry Würzburg, FRG, March 24–25, 1988 proceedings. In: Conference Proceedings CG (3rd edn., p. 68).
57. Posdamer, J. L., (1977). A vector development of the fundamentals of computational geometry. Computer Graphics and Image Processing, 6(4), 382–393.
58. Poulos, M., Deliyannis, I., & Floros, A., (2012). Audio fingerprint extraction using an adapted computational geometry algorithm. Computer and Information Science, 5(6), 88.
59. Preparata, F. P., & Shamos, M. I., (1985). Convex hulls: Basic algorithms. In: Computational Geometry (Vol. 3, pp. 95–149). Springer, New York, NY.
60. Reif, J. H., & Sen, S., (1989). Polling: A new randomized sampling technique for computational geometry. In: Proceedings of the 21st Annual ACM Symposium on Theory of Computing (2nd edn., pp. 394–404).
61. Salesin, D., Stolfi, J., & Guibas, L., (1989). Epsilon geometry: Building robust algorithms from imprecise computations. In: Proceedings of the Fifth Annual Symposium on Computational Geometry (Vol. 2, pp. 208–217).

62. Samuel, G. L., & Shunmugam, M. S., (2000). Evaluation of circularity from coordinate and form data using computational geometric techniques. Precision Engineering, 24(3), 251–263.
63. Saračević, M. H., Stanimirović, P. S., & Mašović, S. H., (2013). Object-oriented analysis and design for one algorithm of computational geometry: Forward, reverse and round-trip engineering. JITA: Journal of Information Technology and Applications, 6(2), 1–5.
64. Schirra, S., (1996). Precision and robustness in geometric computations. In: Advanced School on the Algorithmic Foundations of Geographic Information Systems (pp. 255–287). Springer, Berlin, Heidelberg.
65. Schirra, S., (1998). Robustness and Precision Issues in Geometric Computation, 7(5), 3–6.
66. Schorn, P., (1991). Robust Algorithms in a Program Library for Geometric Computation (Vol. 4, No. 5, pp. 4–8). Doctoral dissertation, ETH Zurich.
67. Schurmann, A., (2009). Computational Geometry of Positive Definite Quadratic Forms: Polyhedral Reduction Theories, Algorithms, and Applications (Vol. 48, pp. 1–4). American Mathematical Soc.
68. Šeda, M., & Seda, P., (2020). Computational geometry data structures in logistics and navigation tasks. In: 2020 12th International Congress on Ultra Modern Telecommunications and Control Systems and Workshops (ICUMT) (Vol. 2, pp. 179–184). IEEE.
69. Sen, S., (1997). Lower bounds for parallel algebraic decision trees, parallel complexity of convex hulls and related problems. Theoretical Computer Science, 188(1, 2), 59–78.
70. Sethian, J. A., (1999). Level Set Methods and Fast Marching Methods: Evolving Interfaces in Computational Geometry, Fluid Mechanics, Computer Vision, and Materials Science (Vol. 3, pp. 1–6). Cambridge university press.
71. Sugihara, K., (2000). Three-dimensional convex hull as a fruitful source of diagrams. Theoretical Computer Science, 235(2), 325–337.
72. Sugihara, K., Iri, M., Inagaki, H., & Imai, T., (2000). Topology-oriented implementation—An approach to robust geometric algorithms. Algorithmica, 27(1), 5–20.
73. Tamassia, R., (1996). Strategic directions in computational geometry. ACM Computing Surveys (CSUR), 28(4), 591–606.
74. Tang, M., Zhao, J. Y., Tong, R. F., & Manocha, D., (2012). GPU accelerated convex hull computation. Computers & Graphics, 36(5), 498–506.
75. Torr, P. H., Zisserman, A., & Maybank, S. J., (1998). Robust detection of degenerate configurations while estimating the fundamental matrix. Computer Vision and Image Understanding, 71(3), 312–333.

76. Toussaint, G., (1994). Advances in computational geometry for document analysis. In: Proc., 3rd Annual Symp. Document Analysis and Information Retrieval, 3(2), 1–5.
77. Wan, F., Shang, H., Wang, L. X., & Sun, Y. X., (2005). How to determine the minimum number of fuzzy rules to achieve given accuracy: A computational geometric approach to SISO case. Fuzzy Sets and Systems, 150(2), 199–209.
78. Wu, C., Wang, X., Bai, D., & Zhang, H., (2008). A fast training algorithm for SVM based on the convex hulls algorithm. In: 2008 9th International Conference on Signal Processing (Vol. 1, pp. 1578–1581). IEEE.
79. Yap, C. K., (1997). Towards exact geometric computation. Computational Geometry, 7(1, 2), 3–23.

CHAPTER

2 Polygon Triangulation

Contents

2.1. INTRODUCTION

Not only the works of great artists are popular with art enthusiasts, but also with criminals. They are incredibly precious, easy to carry, and appear to be rather simple to sell. As a result, art galleries must exercise extreme caution with their holdings (Figure 2.1) (Narkhede and Manocha, 1995).

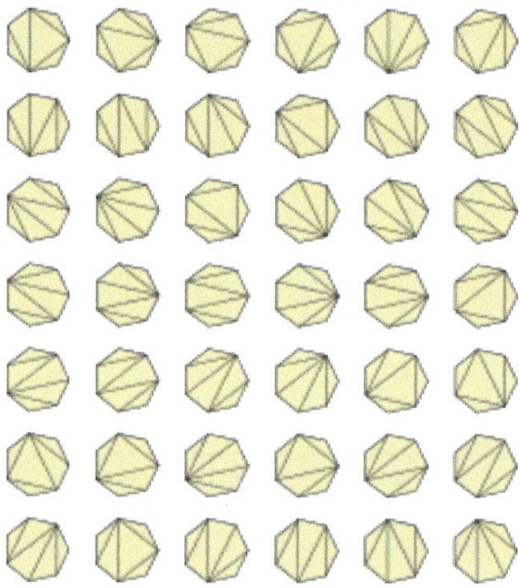

Figure 2.1. Simple diagram showing polygon triangulation.

Source: https://en.wikipedia.org/wiki/Polygon_triangulation.

During the day, attendants may keep an eye on things, but at night, video cameras are required. These cameras are often suspended from the ceiling and spin vertically. The footage to the cams is directed to television displays in the night watchman's office. Because it is simpler to monitor a few TV displays than a large number, the amount of cams should be kept to a minimum. Additionally, a minimal quantity of cameras reduces expenses of a safety system (Goodrich, 1995).

Contrary to this, we cannot take very less cams, as each section of a gallery has to get viewable to at least one. As a result, we should arrange the cameras strategically like that everyone protects a significant portion of the gallery. This results in the establishment of what is frequently known to as Art Gallery. The issue is this: how so many cams do we require to guard a particular gallery, and how do we select where to install them (Kirkpatrick et al., 1992)?

2.2. GUARDING AND TRIANGULATIONS

To more accurately characterize the art gallery issue, we must primarily codify a concept of gallery. Of course, a gallery is a three-dimensional environment, but a ground strategy provides sufficient data to position the cams. As a result, we represent a gallery as a planar polygonal zone. We also limit ourselves to simple polygonal areas, that is, regions consisting of single closed polygonal chain which does not overlap on its own. As a result, we do not permit areas with openings (Bose et al., 1997). In the gallery, each camera position according to the point on the polygon. A cam views such points in the polygon with that it can link with an open area in the polygon's interior.

How many cameras are required to protect the basic polygon? This obviously relies on the polygon in question: the more complicated the polygon, the more cams are necessary. As a result, we will describe the upper constraint on the amount of cams required in the form of n, the quantity of polygon vertices (Bajuelos et al., 2015). Even if two polygons have the similar number of vertices, one may be simpler to protect than the other. A convex polygon, for instance, can be protected by a single camera. To be safe, we will consider the worst condition, which is, we will provide a limit that is valid for every easy polygon with 'n' vertices. (It would be wonderful if we can identify the smallest amount of cameras for the precise polygon we've been given, rather than simply a worst-case bound.) Inappropriately, determining the minimal amount of cameras for a particular polygon is an NP-hard task (Figures 2.2 and 2.3) (Zarei and Ghodsi, 2007).

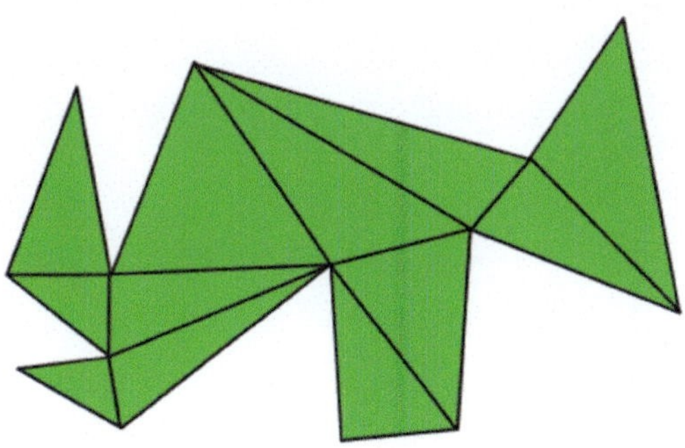

Figure 2.2. Illustration of polygon guarding.

Source: https://slideplayer.com/slide/4832340/.

Figure 2.3. Artwork of polygon triangulation.

Source: https://www.pinterest.com/pin/302515299969919963/.

Assume that 'P' is a basic polygon with 'n' vertices. Since 'P' has a strange form, that's impossible to determine how many cameras we'll need to protect it. As a result, we first deconstruct 'P' into easy-to-protect components, specifically triangles. It is accomplished by creating diagonals among couples of vertices (Figure 2.4) (Hoffmann and Kriegel, 1996).

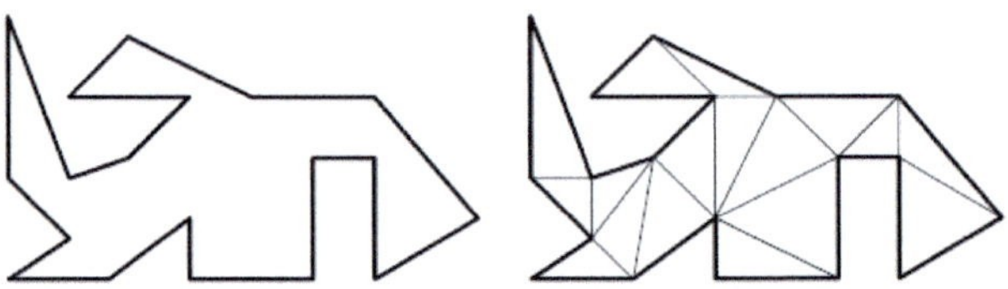

Figure 2.4. A simple polygon and a possible triangulation of it.

Source: https://www.slideshare.net/kasunrangawijeweera/accurate-simple-and-efficient-triangulation-of-a-polygon-by-ear-removal-with-lowest-memory-consumption.

A diagonal is an unbound line path that exists in the inside of 'P' and links two vertices of P. A triangulation of the polygon is a breakdown of a polygon into triangles by a maximum list of non-diagonals. (To guarantee that no triangle contains a polygon vertex in an inner of one of its corners, we demand that the collection of no connecting diagonals be maximum (Everett and Rivera-Campo, 1997). This is possible if the polygon contains three straight collinear vertices.) Triangulations are not frequently distinctive;

for example, the polygon in Figure 2.4 can be triangulated in a variety of methods. We could protect 'P' by deploying a camera in each triangle of P's triangulation TP. Is a triangulation always existing? And how much triangles can a triangulation contain? The next theorem provides solutions to these problems (Canales et al., 2015).

- **Theorem 2.1:** Any easy polygon with 'n' vertices permits a triangulation, and any triangulation of an easy polygon with 'n' vertices comprises of precisely n_2 triangles.

Proof: Induction on 'n' is used to prove this theorem. When n = 3, the polygon is a triangle, and the theorem is obvious. Assume n > 3 and which the theory holds correct for every m n. Assume 'P' is a polygon with 'n' vertices (Bhattacharya et al., 2001). We begin by demonstrating the presence of a diagonal in P. Suppose 'v' be P's leftmost vertex. (In the event of a tie, the lower leftmost vertex is used.) Suppose 'u' and 'w' be the two vertices of 'v' on the border of 'P' that are next to each other. We have discovered a diagonal if the opening section (uw) is in the inner of P. Otherwise, one or extra vertices are located inner to the triangle formed by u, v, and w, or on the diagonal (uw). Letting 'v' be the one furthest away to the line via 'u' and 'w.' The section linking 'v' to 'v' cannot meet an edge of 'P' since this an edge would have an ending within the triangle that is further from line via 'u' and 'w,' which would contradict the previous statement (Addario-Berry et al., 2008).

V as defined by v. As a result, vv is a diagonal.

As a result, a diagonal occurs. 'P' is divided into two easy subpolygons, P_1 and P_2. Assume m_1 represents the quantity of vertices in P_1 and m_2 represents the quantity of vertices in P_2. All m_1 and m_2 must be less than n, therefore P_1 and P_2 may be triangulated by induction. As a result, 'P' may also be triangulated (Karavelas, 2011).

It is still necessary to demonstrate that every triangulation of 'P' comprises of n_2 triangles. Take a random diagonal in a triangulation TP to this end. This diagonal divides 'P' into two subpolygons with vertices m_1 and m_2, correspondingly. Excluding for the vertices forming the diagonal, that appear in either subpolygons, each vertex of 'P' happens in precisely one out of two subpolygons (Speckmann and Tóth, 2005). As a result, $m_1 + m_2 = n + 2$. By induction, each triangulation of Pi has mi 2 triangles, therefore that TP contains $(m_{12}) + (m_{22}) = n_2$ triangles.

- **Theorem 2.2:** Any easy polygon with 'n' vertices may be protected by n_2 cameras. However, putting a camera inner to each triangle looks excessive. A camera positioned on a diagonal, for instance, will protect two triangles; hence, by arranging the cameras on good choice diagonals, we may be capable of limiting the quantity of cameras to around n/2. Cams placed at vertices appear to be even well since a vertex might be happening to multiple triangles, and a camera at that vertex protects all of these. It offers the next strategy (Jungeblut and Ueckerdt, 2020).

Consider T_P to be a triangulation of P. Choose a subset of P's vertices so any triangle in TP has at minimum one of the chosen vertices, then position the cams at the chosen vertices. To discover like a subset, we give a color to every vertex of P: white, gray, or black. Any two vertices linked by an edge or a diagonal will have various colors due to the coloring. This is known as three-coloring a triangulated polygon.

Each triangle in a triangulated polygon with three colors has a white, a gray, and a black vertex. As a result, if we deploy cameras at all gray vertices, for example, we have protected the complete polygon. We can protect 'P' with the most [n/3] cameras by placing them in the smallest color class (Aloupis et al., 2014).

But is there usually a three-coloring? Yes, the answer is yes. We may observe this by looking at what is known as the T_P dual graph. Every triangle in TP has a node in this graph $G(T_P)$. The triangle relating to a node 'v' is meant by t(v). If t(v) and t(μ) share a diagonal, there is an arc among them. $G(T_P)$'s match to T_P's diagonals. Because any diagonal divides 'P' in two, removing one edge from $G(T_P)$ divides the graph in two. As a result, $G(T_P)$ is a tree (Narasimhan, 1997). (Note that this does not apply to a polygon containing holes.) This means we can locate a three-coloring by means of a basic graph traversal, including this type of depth-first search. Then we'll go over how to accomplish it.

We keep the following invariant while doing the depth-first search: all vertices of previously known triangles have been colored white, gray, or black, and no two linked vertices have gotten a very similar color. When all triangles are encountered, the invariant indicates that we have calculated a correct 3-coloring (Michael and Pinciu, 2003). The depth-first search can be initiated from any node of G(TP); the associated triangle's three vertices are colored white, gray, and black. Assume we are traveling from node to a node in G. As a result, t(v) and t(μ) share a diagonal. Because the vertices of t(μ) have been colored, there is only one vertex of t(v) that has to be colored. There is just one color remaining for this vertex, which is the color not utilized for the vertices on the diagonal among t() and t(). Because $G(T_p)$ is a tree, the nodes close to have not yet been stay in, and we may still provide the vertex the leftover color (Bose et al., 2003).

We demonstrate as a basic triangulated polygon can be three-colored. Consequently, of the [n/3] prongs, any basic polygon may be defended with cameras. Nonetheless maybe we must do much better. So, a camera positioned at a vertex may be responsible for more than simply the incident triangles. Inappropriately, there exist easy polygons that need [n/3] cameras for any n. A comb-shaped polygon with a long flat base edge and [n/3] prongs with two edges each is an example (Kaučič et al., 2003). Horizontal edges link the prongs together. The architecture may be designed so that no point in the polygon allows a camera to view into both prongs of the comb at the same time. As a result, we cannot expect for a technique that consistently yields fewer than [n/3] cameras. On the other way in the worst-case scenario, the 3-coloring strategy is best (Liu et al., 1988).

We just established the Art Gallery Theorem, a well-known combinatorial geometry conclusion.

- **Theorem 2.3: (Art Gallery Theorem):** For a basic polygon with 'n' vertices, [n/3] cams are sometimes required but always enough to make each point in the polygon viewable of at least one of the cameras (Keil and Sack, 1985).

We understand that [n/3] cameras are continuously adequate. However, we do not yet have a well-organized technique for computing camera locations. We require a quick algorithm for triangulating a basic polygon. The method should provide an appropriate presentation of the triangulation—say, a doubly-connected edge list—so that we may walk from a triangle to its neighbors in constant time (O'Rourke et al., 2017). With a such description, we can use the approach given above to calculate a collection of at most [n/3] camera locations in linear time: Use a depth-first analysis on the dual graph to calculate a 3-coloring and place the cameras in the lowest color class. In the following sections, we will show you how to do a triangulation in O(nlogn) time. We have previously stated the ultimate outcome about defending a polygon in anticipation of this (Aichholzer et al., 1996).

- **Theorem 2.4:** Assume that 'P' is a basic polygon with 'n' vertices. In O(nlogn) time, a collection of camera placements in 'P' such that each point within 'P' is viewable of at least one of the cameras can be calculated.

2.3. PARTITIONING A POLYGON INTO MONOTONE PIECES

Suppose 'P' be a basic n-vertices polygon. A triangulation of 'P' always exists, as we observed in Theorem 2.1. The proof is positive, and its mains to a recursive triangulation algorithm: locate a diagonal and recursively triangulate the two subpolygons that arise. To identify the diagonal, start with P's leftmost vertices and link it to its two neighbors, u, and w; if that fails, link 'v' to the vertex the furthest from uw within the triangle created by u, v, and 'w' (Yap, 1988). Finding a diagonal takes linear time this method. 'P' may be divided into a triangle and a polygon with n_1 vertices by this diagonal. This has always been the case if we are successful in connecting 'u' and 'w.' As a result, in the worst-case scenario, the triangulation procedure will require quadratic time. Is it possible for us to do better? We can certainly do so for various polygon classes (Agarwal and Suri, 1998). Convex polygons, for example, are simple: Draw diagonals from one of the polygon's vertices to all of the other vertices excluding the neighbors. This simply takes linear time. Decomposing 'P' into convex bits and then triangulating the components might be one way to triangulate a non-convex polygon. Unfortunately, partitioning a polygon into convex sections is just as difficult as triangulating it. As a result, we'll deconstruct 'P' into so-called monotone components, that is much easier (Sack and Toussaint, 1988).

An easy polygon is said to be monotone with regard to a line if any line perpendicular to the polygon's connection is linked. To put it another way, the junction should be either a line, a point, or an empty space. The term y-monotone refers to a polygon that is monotone along the y-axis. The following feature of y-monotone polygons is that if we gait from the topmost to the lowest vertex all with left (or right) boundary chain, we constantly go down or horizontal, not ever upward (Berg et al., 2000).

To triangulate the polygon P, we divide it into y-monotone parts first, and then triangulate the parts. As seen below, we can divide a polygon into monotone sections. Suppose going along the left or right boundary chain from the top vertex of 'P' to the lowermost vertex. A turn vertex is a point when the direction of our walking changes from down to up or up to down way. We must get eliminate these turn vertices in order to divide 'P' into y-monotone pieces (Nahar and Sahni, 1988). This may be accomplished by combining diagonals. If both incident edges travel down at a turn vertex 'v' and the inner of the polygon is locally above v, we must find a diagonal that goes up from v. The polygon is divided in two by the diagonal. In both components, the vertex 'v' will appear. Furthermore, 'v' has an edge that goes down (on the initial edge of P) and a corner that goes up in both parts (the diagonal). As a result, 'v' can no longer be a turn vertex in any of them. We must find a diagonal that goes down if both incident edges of a turn vertex move up and the inner nearby resides underneath it. There appear to be many sorts of turn vertices. Now go on a little more specific here (Kavraki and Kolountzakis, 1995; Li et al., 2007).

If we wish to accurately characterize the various sorts of turn vertices, we should focus on vertices with the similar y-coordinate. This is accomplished by defining "below" and "above" as follows: a point 'p' is below another point 'q' if p_y q_y or $p_y = q_y$ and $p_x > q_x$, and 'p' is above 'q' if $p_y > q_y$ or $p_y = q_y$ and p_x q_x. (Imagine rotating the plane slightly clockwise in regard to the axis system, so that no two points contain the similar y-coordinate; the below/above relative we just described is the similar in this slightly rotated plane (Edelsbrunner et al., 1984).

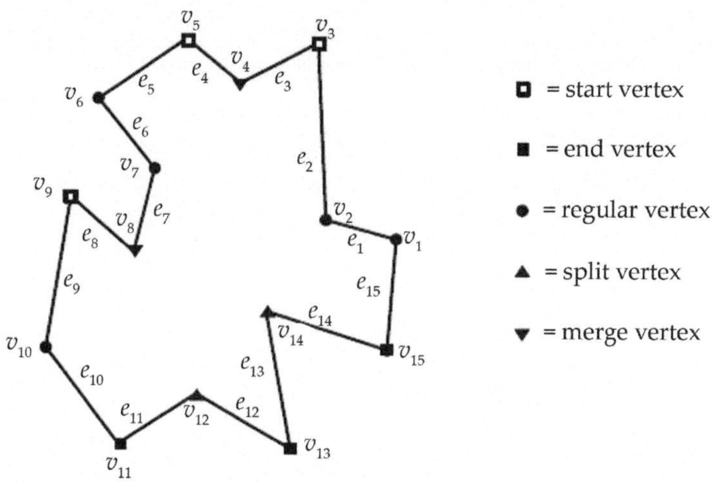

Figure 2.5. Five different sorts of vertices shown.

Source: https://algs4.cs.princeton.edu/41graph/.

In P, there are five different sorts of vertices (see Figure 2.5). Begin vertices, divide vertices, finish vertices, and combine vertices are the four types of turn vertices. The following is a list of their definitions (Park and Shin, 2002). If its two neighbors are under it and the internal angle at 'v' is smaller than, it is a start vertex; if the interior angle is more than, it is a split vertex. (The inner angle cannot be accurate if both neighbors are below v.) If two neighbors are upwards and the interior angle at 'v' is smaller than, the vertex is an end vertex; if the interior angle is bigger than, the vertex is a merge vertex (Krohn and Nilsson, 2013). Regular vertices are those that aren't turn vertices. As a result, a regular vertex has one neighbor above it and the other below it. Because the method would perform a descending plane sweep while keeping the overlap of the sweep line with the polygon, these names were chosen. A part of the intersection divides whenever the sweep line hits a split vertex, two components combine when it meets a merge vertex, and so on (Kavraki et al., 1993).

Local non-monotonicity is caused by the divide and combine vertices. Even more so, the following assertion is correct.

- **Theorem 2.5:** If a polygon contains no split or merge vertices, it is said to be y-monotone.

Proof: Suppose 'P' isn't a y-monotone. We must demonstrate that 'P' has a split or merge vertex (Toussaint and Gindy, 1984).

There is a straight line which crosses 'P' in much more than one linked component because 'P' is not monotone. We may make the very left component a section rather than a single point. Suppose 'p' be the left end and 'q' be the right end of this section (Bärtschi and Suri, 2014). We start at 'q' and follow the border of 'P' to the left of the edge. (This indicates we start at 'q' and work our way up.) The boundary will meet again at a few points, let's call it r. If r = p, as shown in Figure 2.6(a), the tallest vertex met on the way since 'q' to 'r' must be a splitting vertex, and we are finished (Dwivedi and Kovacevic, 2004).

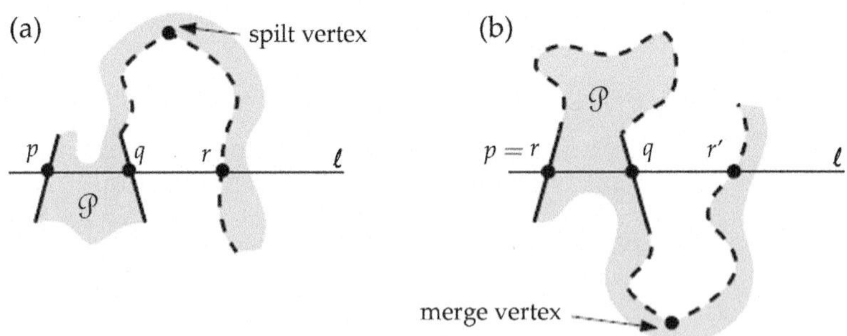

Figure 2.6. Two situations in the demonstration of Theorem 3.4 are shown.

Source: https://link.springer.com/book/10.1007/978-3-540-77974-2.

If r = p, as shown in Figure 2.6(b), we start at 'q' and track the border of 'P' in the opposite direction. The border will cross again, like before.

Suppose 'r' be the point at which this occurs. We can't have r = p since it will suggest P's border crosses just twice, which contradicts the fact that 'P' intersects in several components. As a result, we contain r p, meaning as the smallest vertex met on the way from 'q' must be a merging vertex (Aichholzer et al., 2018).

P has now been divided into y-monotone segments, according to Theorem 2.4, once the split and merge vertices have been removed. We achieve that through extending a diagonal from every split vertex upward and a diagonal from every merging vertex below. Of course, these diagonals should not cross one other. 'P' has now been divided into y-monotone portions as a result of this (Böhringer et al., 1999).

Let's look at how to add the diagonals to the divided vertices first. For this, we employ the plane sweep approach. An anti-clockwise enumeration of the vertex of 'P' is v_1, v_2, …, v_n. Suppose e_1, …, e_n be the set of P's edges, with $e_i = v_i v_{i+1}$ for 1 i < n and e_n = . An imagined sweep line is moved downward above the plane via the plane sweep algorithm. At specific event points, the sweep line comes to a standstill. Such will be the vertex of 'P' in our scenario; throughout the sweep, no additional event points will be formed. The event points are kept in a queue called 'Q' (Wei et al., 2012). The event queue is a precedence queue, with a vertex's y-coordinate determining its priority. If two vertices have the similar y-coordinate, the one on the left takes precedence. The next event to handle may be determined in O(logn) time this manner. (Since no novel events are created throughout the sweeps, we can sort the vertices on their y-coordinates beforehand the sweep and use the organized list to identify the next event in O(1) time (Elnagar and Lulu, 2004). The sweep's purpose is to connect every split vertex to a vertex upon it by adding diagonals. Assume the sweep line comes to a split vertex v_i. Should we link v_i to that vertex? A vertex near v_i is a suitable possibility since we can presumably link v_i to this vertex in the absence of overlapping any of P's edges. Let's go a little more specific here (Lopez and Mehta, 1996). Suppose e_j be the sweep line edge directly to the left of v_i, and e_k be the sweep line edge directly to the right of v_i. Then, among e_j and e_k, and over v_i, we may always link v_i to the smallest vertex. We can link v_i to the higher terminus of ej or the higher endpoint of e_k there is no like that a vertex. This vertex is known as e_j's helper and is denoted by helper (e_j). In formal terms, helper(e_j) is smallest vertex just upon sweep line where the horizontal section linking the vertex to e_j is into P. It's worth noting that helper(e_j) can be e_j's higher endpoint (Heffernan, 1993).

We now understand how to avoid split vertices: join them to the edge to their left's helper. How about combining vertices? They appear to be harder as to ignore since they required a diagonal to a lower vertex than they are. We can't inset such a diagonal when we hit a merging vertex since the section of 'P' below the sweep line hasn't been investigated yet (Lien and Amato, 2006). Fortunately, this issue is simpler than it appears at first glance. Let's say the sweep line comes to a merging vertex v_i. Suppose ej and ek be the endpoints on the sweep line directly to the right and left of v_i, respectively. When we get there, notice how v_i becomes e_j's new assistant. Between e_j and e_k, we want to link v_i to the tallest vertex underneath the sweep line (Agarwal et al., 2021). This is the polar reverse of what we accomplished with split vertices, when the smallest vertex just above sweep line among e_j and e_k was linked. Merge vertices are divided vertices turned upside down, which is not surprising. Of course, when we reach v_i, we have no idea what the topmost vertex underneath the sweep line is.

But it'll be simple to locate later: when we get to a vertex v_m that substitutes v_i as e_j's helper, that's the one we're searching for (Halperin and Wilson, 1994). As a result, whenever we update an edge's helper, we check to see if the previous helper is a merge vertex, and if it is, we add the diagonal among the old and new helpers. Whenever the new helper is a split vertex, this diagonal is always applied to eliminate the split vertex. We get away of a split vertex and a merge vertex with the similar diagonal if the old helper was a merge vertex. It's also possible that under v_i, the e_j helper isn't updated. We can link v_i to e_j's lower endpoint in this scenario (Chen et al., 2004).

We need to locate the boundary to the left of every vertex in the technique above. As a result, in the leaves of a dynamic binary search tree T, we collect the edges of 'P' crossing the sweep line. The edges' left-to-right order matches to the left-to-right order of T's leaves (Bose et al., 2009). We only require to keep edges in 'T' that have the interior of 'P' to their right since we're only concerned in corners to the left of split and merge vertices. We keep track of each edge's helper in T.

The sweep line algorithm's state is determined by the tree 'T' and the helpers placed with the edges. As the sweep line travels, the state changes: edges begin or stop crossing the sweep line, and an edge's helper may be replaced. 'P' is divided into subpolygons that must be processed by the algorithm at a later time (Aichholzer et al., 1995). We'll go into more detail. We will save the subdivision caused by 'P' and the extra diagonal lines in a doubly-connected edge list 'D' to have quick access to these subpolygons. We suppose that 'P' is initially stated as a doubly-connected edged list; if 'P' is specified in another way—for example, as an anti-clockwise collection of its vertices—we must first create a doubly-connected edge list for 'P' (Gewali et al., 1995). The doubly-connected edge list is updated using the diagonal lines computed for the split and merge vertices. We employ cross-pointers among the edges corners in the state structure and the equivalent vertices in the doubly-connected edge list to reach the doubly-connected edge list. With some basic pointer manipulations, adding a diagonal may be done in real-time. The following is the current global algorithm (Dumitrescu and Pach, 2002).

- **MAKEMONOTONE is an algorithm (P):**

Input: In a doubly-connected side list D, a simple polygon 'P' is kept.

Output: 'P' is partitioned into monotone subpolygons, which are kept in D.

Create a priority queue 'Q' on the vertices of P, with their y-coordinates serving as the priority. When two points with the same y-coordinate have the same x-coordinate, the one with the lower x-coordinate gets the greater priority (Edelsbrunner et al., 1986).

- Create a blank binary search tree T. Initialize an empty binary search tree T;
- If 'Q' is not empty;
- Delete the vertex v_i from 'Q' that has the highest priority;
- Depending on the kind of vertex, call the relevant operation.

Following that, we'll go through how to manage the event points in further detail. You should examine these methods first without considering degenerate instances, and then verify that they are accurate in degenerate circumstances later (Garey et al., 1978). (To this aim, in line 1 of HANDLESPLITVERTEX and line 2 of HANDLEMERGEVERTEX, you should define "immediately left of" appropriately.) When dealing with a vertex, we must always perform two things. First, we must determine if a diagonal is required. This is true for a split vertex as well as for replacing an edge's helper when the prior helper was a merge vertex. Second, the data in the status structure 'T' must be updated. Below are the exact algorithms for each sort of occurrence. You can see what occurs in each of the scenarios by looking at the sample figure on the following page (Wagener, 1988).

HANDLESTARTVERTEX(v_i)
1. Insert e_i in $\mathcal{T}$ and set $helper(e_i)$ to v_i.

At the start vertex v_5 in the example figure, for instance, we insert e_5 into the tree $\mathcal{T}$.

HANDLEENDVERTEX(v_i)
1. **if** $helper(e_{i-1})$ is a merge vertex
2. **then** Insert the diagonal connecting v_i to $helper(e_{i-1})$ in $\mathcal{D}$.
3. Delete e_{i-1} from $\mathcal{T}$.

Whenever we approach end vertex v_{15} in the running case the edge e_{14}'s helper is v_{14}. We don't require to add a diagonal because v_{14} is not a merging vertex (Fournier and Montuno, 1984).

- **HANDLESPLITVERTEX(v_i):**
  - In T, locate a line ej straight to the left of v_i;
  - In D, add the diagonal that connects v_i to helper(e_j);
  - Assister(ej) vi;
  - Replace e_i in 'T' with v_i and set helper(e_i) to v_i.

In our case, e_9 is the side to the left for split vertex v_{14}. Because v_8 is its assistant, we include diagonally from v_{14} to v_8 (Narkhede and Manocha, 1995).

- **HANDLEMERGEVERTEX(v_i):**
  - If assist(e_i – 1) is a merge vertex;
  - After that, in D, add the diagonal that connects v_i to helper (e_i1);
  - Subtract e_i – 1 to T;
  - In T, locate the edge e_j just to the left of v_i;
  - The case where helper(e_j) is a merge vertex;
  - Then, in D, add the diagonal bridging v_i and helper(e_j);
  - Assister(ej) v_i.

In this case, the merging vertex v_8 is a helper vertex, thus we insert a diagonal from v_8 to v_2.

The only remaining function to discuss is the one for dealing with a normal vertex. The activities that must be taken at a normal vertex are determined by whether 'P' is located to the vertex's left or right (Toussaint, 1984).

- **HANDLEREGULARVERTEX(v_i):**
  - If P's interior is to right of v_i;

- If helper(e_i – 1) is a merge vertex;
- Then, in D, add the diagonal that connects v_i to helper(e_i – 1);
- Subtract e_i – 1 from T;
- Replace e_i in 'T' with v_i and set helper(e_i) to v_i;
- Else in T, look for the line e_j directly to the left of v_i;
- In the case where helper(e_j) is a merge vertex;
- Then, in D, insert the diagonal that connects v_i to helper(e_j);
- Assister(ej) vi.

For instance, in instance, we include a diagonal through v_6 to v_4 at the ordinary vertex v_6. It keeps to demonstrate that MAKEMONOTONE splits 'P' properly into monotone bits (Goodrich, 1989).

2.4. TRIANGULATING A MONOTONE POLYGON

We just saw in which way a basic polygon divides into y-monotone parts in O(nlogn) time. This isn't that intriguing in and of itself. However, we prove in this part that monotone polygons may be triangulated in constant time. These findings suggest that every easy polygon may be triangulated in O(nlogn) time, which is a noticeable difference over the quadratic time technique outlined at the start of the preceding part (Yap, 1988).

Assume that 'P' is a y-monotone polygon with 'n' vertices. For the time being, we will suppose that 'P' is absolutely y-monotone, that is, that it is y-monotone and does not have any horizontal edges. When we walk from the top vertex of 'P' to the lower vertex of 'P' on the left or right border chain of P, we always travel. This is the trait that allows us to easily triangulate a monotone polygon: we can work our way around 'P' from up to down on together chains, inserting diagonals when available. Following that, we will go into the specifics of this excessive triangulation technique (Toussaint and Avis, 1982).

The methodological controlled the vertices in lessening y-coordinate order. If two vertices have the similar y-coordinate, the one on the left is processed first. As an auxiliary data structure, the technique requires a stack S. The stack is initially blank; afterwards, it stores the vertices of 'P' that have been observed but may still require further diagonals (Bern et al., 1992; Heffernan and Mitchell, 1990). When we grip a vertex, we insert as numerous diagonals as feasible from this vertex to vertices on the stack.

These diagonals divide 'P' into triangles. The vertices that have been treated but not broken off—the vertices on the stack—are on the border of the remaining section of 'P' that has to be triangulated. The bottommost of these vertices, which was encountered last, is at the top of the stack, the lowest average is second on the stack, and so on (Ronfard and Rossignac, 1994). The section of 'P' that still has to be triangulated and is above every vertex encountered so far has a unique shape: it resembles an upside-down funnel. One border of the funnel is made up of a portion of a single edge of P, while the other is a chain made up of reflex vertices, which means that the internal angle at these vertices is at least 180. Only the bottommost vertex, at the top of the stack, is convex. This characteristic holds true even after we've dealt with the following vertex. As a result, it is an algorithm invariant (ElGindy and Toussaint, 1988).

Let's look at the diagonals we can insert whenever we get to the next vertex popped. There are two possibilities: either v_j, the next vertex to be treated, is on the similar chain as the previous vertices in the stack, or it is on the reverse chain (Chazelle and Incerpi, 1984). If v_j is on the opposing chain, it must have been the bottom terminal of the funnel's single edge e. Because of the funnel's form, we may add diagonals from v_j to all of the vertices now on the stack but the final one (the one at the bottom); the very last vertex on the stack is the top vertex of e, thus it is already linked to v_j. All of these vertices have been removed from the stack (Lee and Chwa, 1987). The untriangulated area of the polygon above v_j is restricted by the diagonal that links v_j to the preceding vertex on the header and the border of 'P' popped and continuing downhill from this vertex, creating the appearance of a funnel, and preserving the invariant pushed. Because this vertex and v_j are still part of the untriangulated polygon, they are put onto the stack (Chazelle, 1991; Lingas, 1988).

The other instance is whenever v_j is on the similar chain as the stack's reflex vertices. We may not be capable of drawing diagonals from v_j to all vertices on the stack this time (Asano and Kirkpatrick, 2013). However, because the ones to that we may link v_j are all sequential and at the top of the stack, we can take the following steps (Tarjan and Van, 1988). To begin, remove one vertex from the stack; this vertex has already been linked to v_j through an edge of P. Following, we'll take vertices from the stack and link them to v_j until we reach a point where this isn't possible. By inspecting v_j, v_k, and the preceding vertex that was popped, we can determine if a diagonal may be made from v_j to a vertex v_k on the stack (Lamot and Zalik, 2000). When we come across a vertex to that which we are unable to link v_j, we pop the most recently popped vertex back into the stack. This is either the latest vertex through which a diagonal was inserted or the neighbor of v_j on the border of P. We add v_j to the stack in the following (Figure 2.7) (Lamot and Zalik, 1999).

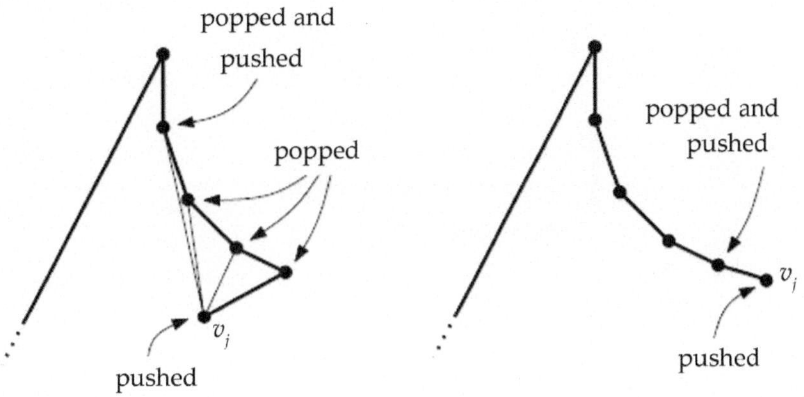

Figure 2.7. Two scenarios when the next vertex is on the similar side as the reflex vertices on the stack.

Source: https://www.sciencedirect.com/science/article/pii/S0166218X01002566.

The invariant is recovered in both cases: one side of the funnel is confined by a portion of a similar edge, while the further side is constrained by a chain of reflex vertices. We obtain the algorithm shown below (Amato et al., 2001; Held, 2001).

- **Algorithm TRIANGULATEMONOTONEPOLYGON(P):**

Input: A polygon 'P' that is strictly y-monotone and is recorded in a doubly-connected edge list D.

Output: In the doubly-connected edge list D, a triangulation of 'P' is kept.

- Combine the vertices on P's left and right chains into a single series, ordered by decreasing y-coordinate. If two vertices have the similar y-coordinate, the one on the left takes precedence (Chin and Wang, 1998). The sorted sequence is denoted by $u_1, \ldots, u_n$;
- Create a blank stack 'S' and place u_1 and u_2 on it;
- For j_3 to n_1;
- Do if u_j and the vertex on top of 'S' are on dissimilar chains;

- Then Pop all vertices from S;
- With the exception of the final popped vertex, insert a diagonal through uj into D;
- Transfer uj_1 and uj to S;
- Else remove one of S's vertices;
- As far even as diagonals through u_j to them and are inside P, release another vertex from S. Fill in the blanks in 'D' with these diagonals. Return the last popped vertex to its original position on 'S' (Chazelle, 1990) ;
- u_j should be pushed onto S;
- Insert diagonals from un to all vertices in the stack except the first and last.

How long would the algorithm start taking? Step 1 requires linear time, whereas Step 2 requires constant time. The for-loop is run n3 times, with each execution taking linear time. However, only two vertices are pushed during each execution of the for-loop (Berkman et al., 1993). As a result, the overall amount of pushes, such as the two in Step 2, is limited to $2n_4$. Since the quantity of pops might not be bigger than the amount of pushes, the overall duration for all for-loop executions is O(n). The final step of the method similarly requires at the most linear time; hence the overall algorithm takes O(n) time (Varshosaz, 2005; Keil and Sack, 1985).

- **Theorem 2.6:** In linear time, a strictly y-monotone polygon with 'n' vertices may be triangulated.

We needed a monotone polygon triangulation technique as a subroutine for triangulating arbitrary basic polygons. The plan was to breakdown a polygon into monotone bits before triangulating these components. We appear to have all of the necessary elements. However, there is one problem: in this part, we believed that the source is an absolutely y-monotone polygon, although the previous section's approach may yield monotone parts with straight edges (Seide, 1991). Remember how we regarded vertices with the similar y-coordinates from left to right in the last section? This has the same effect as a little clockwise rotation of the plane, so that no two vertices are on a straight axis. As a result, the monotone subpolygons generated by the preceding section's technique are absolutely monotone in this slightly rotated plane (Žalik and Lamot, 2000). As a result, the current section's triangulation procedure works appropriately if we consider vertices with the similar y-coordinate as left to right (that is similar to working in the rotated axis). As a result, we may combine the two techniques to create a triangulation algorithm that can be applied to any basic polygon (Chen, 1995; Asano and Umeo, 1987). How long does it take the triangulation algorithm to complete? According to Theorem 2.4, breaking the polygon into monotone sections takes O(nlogn) time. In the second stage, we use the linear-time approach from this section to triangulate each monotone component. The second step takes O(n) time in total because the combination of the quantity of vertices of the components is O(n). As a consequence, we receive the following result (Czumaj, 1993; Lee and Lin, 1986).

- **Theorem 2.7:** A basic polygon with 'n' vertices may be triangulated in O(nlogn) time using an O(n) storage technique.

We've already learned how to triangulate basic polygons. What of polygons with holes? Can they likewise be simply triangulated? Yes, the response is yes. In reality, the approach we've seen operates for polygons with openings: we didn't use the knowledge that the polygon was easy anywhere in the process for splitting it into monotone parts (Kirkpatrick et al., 1992). It also operates in a broader context: Assume we have a planar subdivision 'S' and we wish to triangulate it. More specifically, if 'B' is a bounding box containing all of S's edges in its interior, we want to discover the greatest amount of non-diagonals—line segments linking vertices of 'S' or 'B' that do not overlap S's edges—that divide 'B' into triangles. A triangulated subdivision is seen in Figure 2.8 (Held and Mitchell, 2008). The subdivision and bounding box edges are shown in bold. We may apply the technique from this chapter to compute such a triangulation: first divide the split into monotone parts, and then triangulate the parts. As a result, the following theorem follows (Asano and Umeo, 1988).

- **Theorem 2.8:** A planar subdivision with 'n' total vertices can be triangulated in O(nlogn) time using an O(n) storage technique.

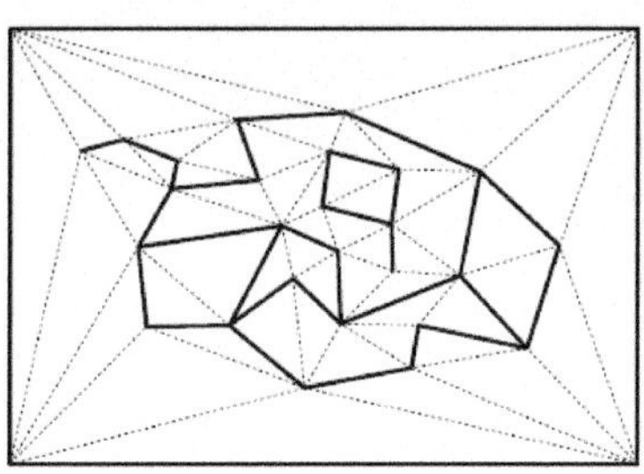

Figure 2.8. A triangulated subdivision.

Source: https://www.pinterest.com/pin/111816003221155839/.

REFERENCES

1. Addario-Berry, L., Amini, O., Sereni, J. S., & Thomassé, S., (2008). Guarding art galleries: The extra cost for sculptures is linear. In: Scandinavian Workshop on Algorithm Theory (Vol. 1, pp. 41–52). Springer, Berlin, Heidelberg.
2. Agarwal, P. K., & Suri, S., (1998). Surface approximation and geometric partitions. SIAM Journal on Computing, 27(4), 1016–1035.
3. Agarwal, P. K., Aronov, B., Geft, T., & Halperin, D., (2021). On two-handed planar assembly partitioning with connectivity constraints. In: Proceedings of the 2021 ACM-SIAM Symposium on Discrete Algorithms (SODA) (Vol. 4, No. 5, pp. 1740–1756). Society for Industrial and Applied Mathematics.
4. Aichholzer, O., Alberts, D., Aurenhammer, F., & Gärtner, B., (1995). Straight skeletons of simple polygons. In: Proc. 4th Internat. Symp. of LIESMARS (pp. 114–124).
5. Aichholzer, O., Aurenhammer, F., Alberts, D., & Gärtner, B., (1996). A novel type of skeleton for polygons. In: J. UCS The Journal of Universal Computer Science (4th edn., pp. 752–761). Springer, Berlin, Heidelberg.
6. Aichholzer, O., Fabila-Monroy, R., Flores-Penaloza, D., Hackl, T., Urrutia, J., & Vogtenhuber, B., (2018). Modem illumination of monotone polygons. Computational Geometry, 68, 101–118.
7. Aloupis, G., Bose, P., Dujmović, V., Gray, C., Langerman, S., & Speckmann, B., (2014). Triangulating and guarding realistic polygons. Computational Geometry, 47(2), 296–306.
8. Amato, N. M., Goodrich, M. T., & Ramos, E. A., (2001). A randomized algorithm for triangulating a simple polygon in linear time. Discrete & Computational Geometry, 26(2), 245–265.
9. Asano, T., & Kirkpatrick, D., (2013). Time-space tradeoffs for all-nearest-larger-neighbors problems. In: Workshop on Algorithms and Data Structures (Vol. 2, pp. 61–72). Springer, Berlin, Heidelberg.
10. Asano, T., & Umeo, H., (1987). Systolic algorithms for computing the visibility polygon and triangulation of a polygonal region. In: Parallel Algorithms and Architectures (pp. 77–85). Springer, Berlin, Heidelberg.
11. Asano, T., & Umeo, H., (1988). Systolic algorithms for computing the visibility polygon and triangulation of a polygonal region. Parallel Computing, 6(2), 209–216.
12. Bajuelos, A. L., Canales, S., Hernández, G., Martins, M., & Matos, I., (2015). Some results on open-edge and open mobile guarding of polygons

and triangulations. The Computer Journal, 58(1), 160–171.

13. Bärtschi, A., & Suri, S., (2014). Conflict-free chromatic art gallery coverage. Algorithmica, 68(1), 265–283.
14. Berg, M. D., Kreveld, M. V., Overmars, M., & Schwarzkopf, O. C., (2000). Polygon triangulation. In: Computational Geometry (2nd edn., pp. 45–61). Springer, Berlin, Heidelberg.
15. Berkman, O., Schieber, B., & Vishkin, U., (1993). Optimal doubly logarithmic parallel algorithms based on finding all nearest smaller values. Journal of Algorithms, 14(3), 344–370.
16. Bern, M., Dobkin, D., & Eppstein, D., (1992). Triangulating polygons without large angles. In: Proceedings of the Eighth Annual Symposium on Computational Geometry (Vol. 1, pp. 222–231).
17. Bhattacharya, B., Mukhopadhyay, A., & Narasimhan, G., (2001). Optimal algorithms for two-guard walkability of simple polygons. In: Workshop on Algorithms and Data Structures (3rd edn., pp. 438–449). Springer, Berlin, Heidelberg.
18. Bjorling-Sachs, I., & Souvaine, D. L., (1995). An efficient algorithm for guard placement in polygons with holes. Discrete & Computational Geometry, 13(1), 77–109.
19. Böhringer, K. F., Donald, B. R., & Halperin, D., (1999). On the area bisectors of a polygon. Discrete & Computational Geometry, 22(2), 269–285.
20. Bose, P., Kirkpatrick, D., & Li, Z., (2003). Worst-case-optimal algorithms for guarding planar graphs and polyhedral surfaces. Computational Geometry, 26(3), 209–219.
21. Bose, P., Morin, P., Smid, M., & Wuhrer, S., (2009). Rotationally monotone polygons. Computational Geometry, 42(5), 471–483.
22. Bose, P., Shermer, T., Toussaint, G., & Zhu, B., (1997). Guarding polyhedral terrains. Computational Geometry, 7(3), 173–185.
23. Canales, S., Hernández, G., Martins, M., & Matos, I., (2015). Distance domination, guarding and covering of maximal outerplanar graphs. Discrete Applied Mathematics, 181, 41–49.
24. Chazelle, B., & Incerpi, J., (1984). Triangulation and shape-complexity. ACM Transactions on Graphics (TOG), 3(2), 135–152.
25. Chazelle, B., (1990). Triangulating a simple polygon in linear time. In: Proceedings [1990] 31st Annual Symposium on Foundations of Computer Science (Vol. 3, pp. 220–230). IEEE.

26. Chazelle, B., (1991). Triangulating a simple polygon in linear time. Discrete & Computational Geometry, 6(3), 485–524.
27. Chen, D. Z., (1995). Efficient geometric algorithms on the EREW PRAM. IEEE Transactions on Parallel and Distributed Systems, 6(1), 41–47.
28. Chen, D. Z., Hu, X. S., Luan, S., Wang, C., & Wu, X., (2004). Geometric algorithms for static leaf sequencing problems in radiation therapy. International Journal of Computational Geometry & Applications, 2, 14(3–5), 311–339.
29. Chin, F., & Wang, C. A., (1998). Finding the constrained Delaunay triangulation and constrained Voronoi diagram of a simple polygon in linear time. SIAM Journal on Computing, 28(2), 471–486.
30. Czumaj, A., (1993). Parallel algorithm for the matrix chain product and the optimal triangulation problems. In: Annual Symposium on Theoretical Aspects of Computer Science (Vol. 2, pp. 294–305). Springer, Berlin, Heidelberg.
31. Dumitrescu, A., & Pach, J., (2002). Partitioning colored point sets into monochromatic parts. International Journal of Computational Geometry & Applications, 12(05), 401–412.
32. Dwivedi, R., & Kovacevic, R., (2004). Automated torch path planning using polygon subdivision for solid freeform fabrication based on welding. Journal of Manufacturing Systems, 23(4), 278–291.
33. Edelsbrunner, H., Guibas, L. J., & Stolfi, J., (1986). Optimal point location in a monotone subdivision. SIAM Journal on Computing, 15(2), 317–340.
34. Edelsbrunner, H., O'Rourke, J., & Welzl, E., (1984). Stationing guards in rectilinear art galleries. Computer Vision, Graphics, and Image Processing, 27(2), 167–176.
35. ElGindy, H., & Toussaint, G., (1988). On triangulating palm polygons in linear time. In: New Trends in Computer Graphics (Vol. 3, pp. 308–317). Springer, Berlin, Heidelberg.
36. Elnagar, A., & Lulu, L., (2004). Guarding polygons with holes for robot motion planning applications. In: 2004 IEEE International Conference on Systems, Man and Cybernetics (IEEE Cat. No. 04CH37583) (Vol. 1, pp. 923–928). IEEE.
37. Everett, H., & Rivera-Campo, E., (1997). Edge guarding polyhedral terrains. Computational Geometry, 7(3), 201–203.
38. Fournier, A., & Montuno, D. Y., (1984). Triangulating simple polygons and equivalent problems. ACM Transactions on Graphics (TOG), 3(2), 153–174.

39. Garey, M. R., Johnson, D. S., Preparata, F. P., & Tarjan, R. E., (1978). Triangulating a simple polygon. Information Processing Letters, 7(4), 175–179.
40. Gewali, L. P., (1995). Recognizing s-star polygons. Pattern Recognition, 28(7), 1019–1032.
41. Goodrich, M. T., (1989). Triangulating a polygon in parallel. Journal of Algorithms, 10(3), 327–351.
42. Goodrich, M. T., (1995). Planar separators and parallel polygon triangulation. Journal of Computer and System Sciences, 51(3), 374–389.
43. Halperin, D., & Wilson, R. H., (1994). Assembly Partitioning with a Constant Number of Translations (No. SAND-94-1819) (Vol. 3, No. 1, pp. 10–12). Sandia National Labs., Albuquerque, NM, United States.
44. Heffernan, P. J., & Mitchell, J. S., (1990). Structured visibility profiles with applications to problems in simple polygons. In: Proceedings of the Sixth Annual Symposium on Computational Geometry (Vol. 3, pp. 53–62).
45. Heffernan, P. J., (1993). Linear-time algorithms for weakly-monotone polygons. Computational Geometry, 3(3), 121–137.
46. Held, M., & Mitchell, J. S., (2008). Triangulating input-constrained planar point sets. Information Processing Letters, 109(1), 54–56.
47. Held, M., (2001). FIST: Fast industrial-strength triangulation of polygons. Algorithmica, 30(4), 563–596.
48. Hoffmann, F., & Kriegel, K., (1996). A graph-coloring result and its consequences for polygon-guarding problems. SIAM Journal on Discrete Mathematics, 9(2), 210–224.
49. Jungeblut, P., & Ueckerdt, T., (2020). Guarding quadrangulations and stacked triangulations with edges. In: International Workshop on Graph-Theoretic Concepts in Computer Science (Vol. 3, pp. 14–26). Springer, Cham.
50. Karavelas, M. I., (2011). Guarding curvilinear art galleries with edge or mobile guards via 2-dominance of triangulation graphs. Computational Geometry, 44(1), 20–51.
51. KauČiČ, B., Žalik, B., & Novak, F., (2003). On the lower bound of edge guards of polyhedral terrains. International Journal of Computer Mathematics, 80(7), 811–814.
52. Kavraki, L. E., & Kolountzakis, M. N., (1995). Partitioning a planar assembly into two connected parts is NP-complete. Information Processing Letters, 55(3), 159–165.

53. Kavraki, L., Latombe, J. C., & Wilson, R. H., (1993). On the complexity of assembly partitioning. Information Processing Letters, 48(5), 229–235.
54. Keil, J. M., & Sack, J. R., (1985). Minimum decompositions of polygonal objects. In: Machine Intelligence and Pattern Recognition (Vol. 2, pp. 197–216). North-Holland.
55. Kirkpatrick, D. G., Klawe, M. M., & Tarjan, R. E., (1992). Polygon triangulation inO (n log logn) time with simple data structures. Discrete & Computational Geometry, 7(4), 329–346.
56. Krohn, E. A., & Nilsson, B. J., (2013). Approximate guarding of monotone and rectilinear polygons. Algorithmica, 66(3), 564–594.
57. Lamot, M., & Balik, B., (1999). An overview of triangulation algorithms for simple polygons. In: 1999 IEEE International Conference on Information Visualization (Cat. No. PR00210) (Vol. 1, pp. 153–158). IEEE.
58. Lamot, M., & Zalik, B., (2000). Algorithms for triangulating simple polygons. In: ITI 2000. Proceedings of the 22nd International Conference on Information Technology Interfaces (Cat. No. 00EX411) (Vol. 2, pp. 429–436). IEEE.
59. Lee, D. T., & Lin, A. K., (1986). Generalized Delaunay triangulation for planar graphs. Discrete & Computational Geometry, 1(3), 201–217.
60. Lee, S. H., & Chwa, K. Y., (1987). A new triangulation-linear class of simple polygons. International Journal of Computer Mathematics, 22(2), 135–147.
61. Li, J., Wang, W., & Wu, E., (2007). Point-in-polygon tests by convex decomposition. Computers & Graphics, 31(4), 636–648.
62. Lien, J. M., & Amato, N. M., (2006). Approximate convex decomposition of polygons. Computational Geometry, 35(1, 2), 100–123.
63. Lingas, A., (1988). Greedy triangulation can be efficiently implemented in the average case. In: International Workshop on Graph-Theoretic Concepts in Computer Science (Vol, 1, pp. 253–261). Springer, Berlin, Heidelberg.
64. Liu, R., & Ntafos, S., (1988). On decomposing polygons into uniformly monotone parts. Information Processing Letters, 27(2), 85–89.
65. Lopez, M. A., & Mehta, D. P., (1996). Efficient decomposition of polygons into L-shapes with application to VLSI layouts. ACM Transactions on Design Automation of Electronic Systems (TODAES), 1(3), 371–395.
66. Michael, T. S., & Pinciu, V., (2003). Art gallery theorems for guarded guards. Computational Geometry, 26(3), 247–258.
67. Nahar, S., & Sahni, S., (1988). Fast algorithm for polygon decomposition. IEEE Transactions on Computer-Aided Design of Integrated Circuits and Systems, 7(4), 473–483.

68. Narasimhan, G., (1997). On Hamiltonian triangulations in simple polygons. In: Workshop on Algorithms and Data Structures (3rd edn., pp. 321–330). Springer, Berlin, Heidelberg.

69. Narkhede, A., & Manocha, D., (1995). Fast polygon triangulation based on seidel's algorithm. In: Graphics Gems V (Vol. 1, 3, No. 2, pp. 394–397). Academic Press.

70. O'Rourke, J., Suri, S., & Tóth, C. D., (2017). Polygons. In: Handbook of Discrete and Computational Geometry (Vol. 3, pp. 787–810). Chapman and Hall/CRC.

71. Park, S. C., & Shin, H., (2002). Polygonal chain intersection. Computers & Graphics, 26(2), 341–350.

72. Ronfard, R. P., & Rossignac, J. R., (1994). Triangulating multiply-connected polygons: A simple, yet efficient algorithm. In: Computer Graphics Forum (Vol. 13, No. 3, pp. 281–292). Edinburgh, UK: Blackwell Science Ltd.

73. Sack, J. R., & Toussaint, G. T., (1988). Guard placement in rectilinear polygons. In: Machine Intelligence and Pattern Recognition (Vol. 6, pp. 153–175). North-Holland.

74. Seidel, R., (1991). A simple and fast incremental randomized algorithm for computing trapezoidal decompositions and for triangulating polygons. Computational Geometry, 1(1), 51–64.

75. Speckmann, B., & Tóth, C. D., (2005). Allocating vertex π-guards in simple polygons via pseudo-triangulations. Discrete & Computational Geometry, 33(2), 345–364.

76. Tarjan, R. E., & Van, W. C. J., (1988). An O(n\log\logn)-time algorithm for triangulating a simple polygon. SIAM Journal on Computing, 17(1), 143–178.

77. Toussaint, G. T., & Avis, D., (1982). On a convex hull algorithm for polygons and its application to triangulation problems. Pattern Recognition, 15(1), 23–29.

78. Toussaint, G. T., & El Gindy, H. A., (1984). Separation of two monotone polygons in linear time. Robotica, 2(4), 215–220.

79. Toussaint, G. T., (1984). A new linear algorithm for triangulating monotone polygons. Pattern Recognition Letters, 2(3), 155–158.

80. Varshosaz, M., Helali, H., & Shojaee, D., (2005). The methods of triangulation. In: Map Middle East'05: Proceedings of the 1st Annual Middle East Conference and Exhibition on Geospatial Information, Technology and Applications, 3(2), 4–7.

81. Wagener, H., (1988). Triangulating a monotone polygon in parallel. In: Workshop on Computational Geometry (3rd edn., pp. 136–147). Springer, Berlin, Heidelberg.
82. Wei, X., Joneja, A., & Mount, D. M., (2012). Optimal uniformly monotone partitioning of polygons with holes. Computer-Aided Design, 44(12), 1235–1252.
83. Yap, C. K., (1988). Parallel triangulation of a polygon in two calls to the trapezoidal map. Algorithmica, 3(1), 279–288.
84. Žalik, B., & Lamot, M., (2000). A contribution to triangulation algorithms for simple polygons. Journal of Computing and Information Technology, 8(4), 319–331.
85. Zarei, A., & Ghodsi, M., (2007). A new algorithm for guarding triangulated irregular networks. The CSI Journal on Computer Science and Engineering, 2(2–4), 11–17.

CHAPTER

3 Fundamentals of Line Segment Intersection

Contents

3.1. INTRODUCTION

Maps are a helpful resource of data while traveling to a new nation. They show you at where tourism destinations are located, as well as the railway lines and roads that will take you there. They also show small lakes, and so on. Regrettably, they could also be a cause of annoyance because it is frequently hard to discover the accurate info: although if you understand the approximate location of a small village, it could still be hard to locate it on a map (Crauser et al., 2001).

Geographic information systems (GIS) divide maps into multiple levels to make them more readable. Each surface is a thematic map, containing only one type of data. Thus, there will be a level for highways, a layer for cities, and a layer for rivers, and so forth. A layer's topic might also be vague (Held, 1997). For instance, there can be the tier for people density, average rainfall, grizzly bear environment, and vegetation.

The geometric data contained in a level can be quite varied: a road map layer may contain pieces of line segments (or perhaps curves), a city layer may have points labeled along city names, as well as a vegetation layer may contain a sector of the map into areas labeled with the kind of vegetation. Thematic maps can be displayed by customers of a GIS. To locate a tiny town, you will choose the level containing cities, avoiding distractions such as river and lake names (Arge et al., 1995).

After spotting the town, you're presumably curious on how to get that point. GIS enables consumers to look at an overlay of multiple maps in this manner (see Figure 3.1). You now can find out how to get through the town using a layer of the route map and the map containing cities. When two or maybe more thematic map levels are displayed in conjunction, the crossings in the layer represent points of interest (Chazelle et al., 1994). For instance, when observing the layer of the road layer and the river layer, it would be beneficial to have clearly marked intersections.

The two maps in this example are essentially channels, and the crossings are points. In other cases, the interconnection of complete regions is of interest. For example, climatic geographers may be interested in locating places with pine forests and yearly precipitation among 1,000 and 1,500 millimeters. These areas are the junctions of vegetation map's "pine forest" region and precipitation map's "1,000–1,500" region (Figure 3.2) (Samet and Webber, 1988).

Figure 3.1. Railroads, cities, rivers, and their overlay in western Canada.

Source: https://link.springer.com/book/10.1007/978-3-540-77974-2.

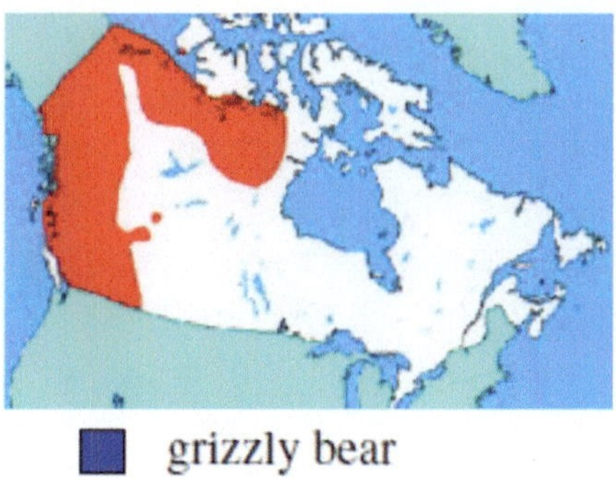

Figure 3.2. Grizzly bear map of Canada.

Source: https://www.canada.ca/en/environment-climate-change/services/species-risk-public-registry/cosewic-assessments-status-reports/grizzly-bear-2012.html.

3.2. LINE SEGMENT INTERSECTION

We begin by looking at the most basic version of the map layers issue, in which the two map levels are symbolized as compilations of parallel lines (Vaishnavi and Wood, 1982). For instance, a map overlay holding small-scale roads, railways, or rivers. It's worth noting that contours can be estimated with a few small segments. The areas caused by these parallel lines will initially be of no interest to us (Figure 3.3) (Chazelle and Edelsbrunner, 1992).

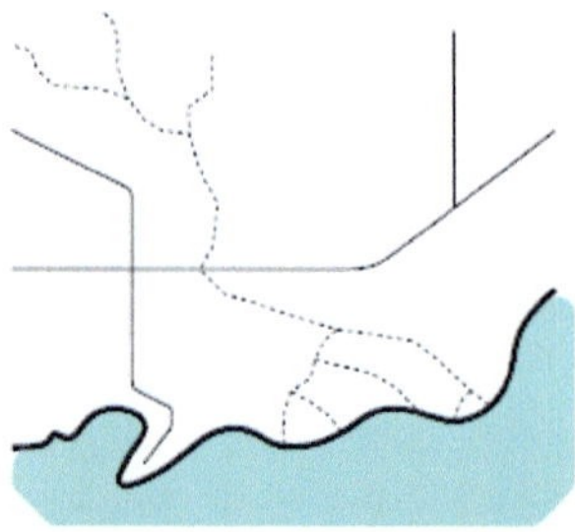

Figure 3.3. Illustration of line segment intersection.

Source: https://people.eecs.berkeley.edu/~jrs/274/.

Far along, we'll see a complicated condition in which the maps aren't just nets, nonetheless explicit segments of the plane into areas. To fix the network overlay issue, we must first define it geometrically. The geometric situation for the base layer of two networks is as follows (Held, 1997; Bartuschka et al., 1997): calculate all interconnections among a section through one collection and a portion from the other given a set of parallel lines. We can't describe when two components overlap in this one stated, so it's not indeed very accurate enough yet.

Is it possible for two segments to cross while one of their endpoints is on the other? To put it another way, we must indicate if the inputs sections are closed or open (Mortensen, 2003). We should return to the program, the network overlay issue, to reach this conclusion.

A passage of a road and a river is represented by inside of one sequence meeting inside of some other chain on a route map and a river intersecting is represented by the inside of one chain meeting the inside of some other chain on a river map (Shamos and Hoey, 1976). This does not rule out the possibility of a point of intersection between the interiors of two parts: the point of intersection might coincide with the termination of a chain segment. In reality, since windy rivers are depicted by numerous smaller sections as well as locations of endings might have been reduced when maps were digitized, this condition is not unusual. We get to the conclusion that the segments should be defined as locked so the node of one section resting on the endpoint of the other segment qualifies as a connection (Figure 3.4) (Arge et al., 1995).

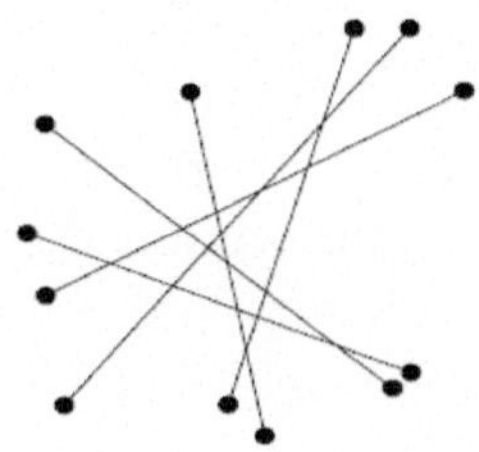

Figure 3.4. Intersecting lines.

Source: https://www.sciencedirect.com/book/9780444825377/handbook-of-computational-geometry.

We'll combine the sections from the two pairs into one set and calculate all intersections between the sections in that setting to simplify the explanation. We'll almost surely locate all of the junctions we're looking for this way (Chazelle et al., 1994; Vahrenhold, 2005). Crossings among sections from the similar set can also be found. Actually, we will, because the sections from one group create a lot of chains in our application, and we consider corresponding nodes as connections. Such additional junctions could be ruled out later by determining if the two sections concerned correspond to the similar set for every reported crossing. As a result, our task is as continues to follow: based on a set 'S' of 'n' enclosed plane sections, report all crossing sites between the sections in 'S' (Figure 3.5) (Saalfeld, 1987).

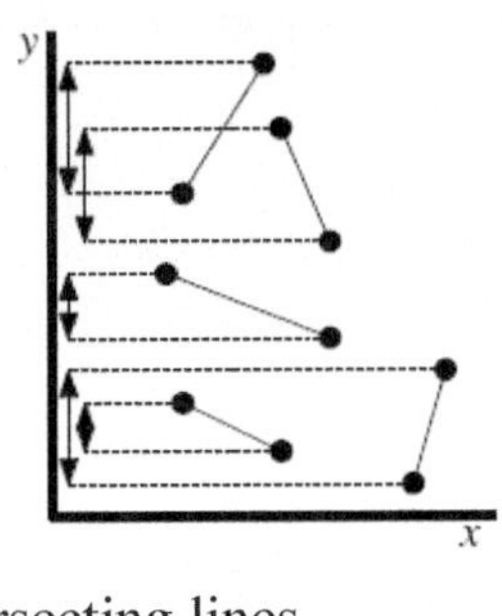

Figure 3.5. Tracking of intersecting lines.

Source: https://www.sciencedirect.com/book/9780444825377/handbook-of-computational-geometry.

This doesn't appear to be a difficult problem: we can just examine every pair of sections, determine if they overlap, and return the junction position if they do. This brute-force approach is plainly $O(n^2)$ in length. In some ways, this is optimal: any algorithm must require $\Omega(n^2)$ time to notify all crossings when every set of segments overlaps. An analogous example may be presented (Forrest, 1991).

When the overlap of two systems is taken into account in practice, however, most segments contact no or just a few other segments, resulting in a far less overall quantity of intersection sites than quadratic. It would be excellent to have a speedier algorithm in these cases. To put it another way, we're looking for an algorithm (Magee and Aggarwal, 1984).

Its execution duration is determined by the number of junction locations as well as the quantity of parts in the input. An output-sensitive algorithm is one in which the running duration of the algorithm is affected by the extent of the output. We may too name like an algorithm intersection-sensitive, because the size of the output is determined by the number of intersections (Ohdan et al., 2014).

How could we ignore performing intersection tests on all sets of segments? In this case, we have to take advantage of the situation's geometry: sections that are near to one another are choices for intersection, whereas sections that are far distant are not. We will demonstrate how to utilize this finding to achieve an output-sensitive algorithm for the horizontal line crossing issue in the sections that follow (Figure 3.6) (Crauser et al., 1990).

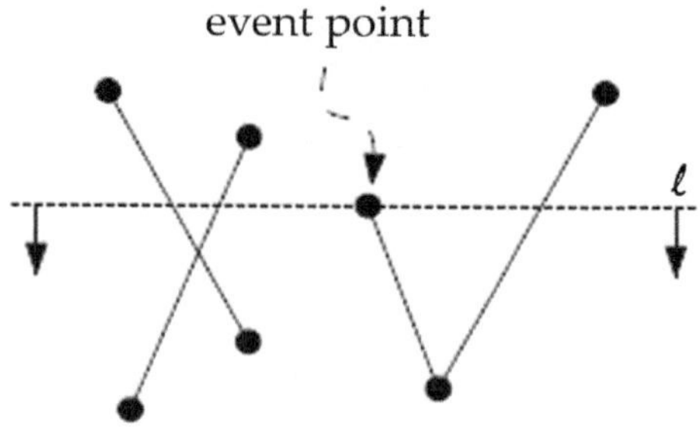

Figure 3.6. Illustration of event points.

Source: https://link.springer.com/book/10.1007/978-3-540-77974-2.

Let $S = s_1, s_2, \ldots, s_n$ denote a collection of portions about whose we wish to compute all interconnections. We would like to ignore testing segments that are too distant from one another.

But how are we going to accomplish it? First, let us just rule out a simple instance. Establish a segment's y-interval as its orthogonal projection onto the y-axis (Rother, 2002). Whenever a set of segments' y-intervals do not overlap—that is, when they are elsewhere in the y-direction—they cannot collide. As a result, we only have to analyze pairings of segments with overlapping y-intervals, i.e., pairs for that a horizontal line cross either segment (Mehlhorn et al., 1993).

Consider sweeping a line downwards across the plane, beginning from a place beyond all segments, to discover these pairings.

We maintain control of all segments interconnecting the imaginary line as we sweep it—details on this would be described later which we can discover the combinations we require (Qi et al., 2017).

This algorithm is referred to as a plane sweep algorithm, and the line is referred to as the event point or sweep line.

The sweep line's status is determined by the pair of sections that intersect it. The status alters intermittently as the sweep line descends (Li et al., 2018). Only at specific points is it necessary to change the status. These points are referred to as the plane sweep algorithm's event points. The event points in this algorithm are the segments' endpoints.

The algorithm only acts when the sweep line attains an event point: it updates the sweep line's status and completes some intersection checks. If an event point is the upper node of the section, a new segment must be added to the status when it begins intersecting the sweep line (Zhou et al., 2009). This segment is compared to those that already intersect the sweep line. If the event point is lower than the sweep line, a segment will no longer intersect the sweep line and should be removed from the status.

This means, we only check sets of segments that are intersected by a horizontal line (Clarkson and Shor, 1989; Hobby, 1999). Unfortunately, this is insufficient: there will still be circumstances in which we test an exponential amount of pairs when there are only a few intersection points. A straightforward illustration is a collection of perpendicular segments that all collide the x-axis. As a result, the algorithm is not output-dependent. The issue would be that two segments intersecting the sweep line can be quite far away horizontally (Figure 3.7) (Bonet and Peraire, 1991).

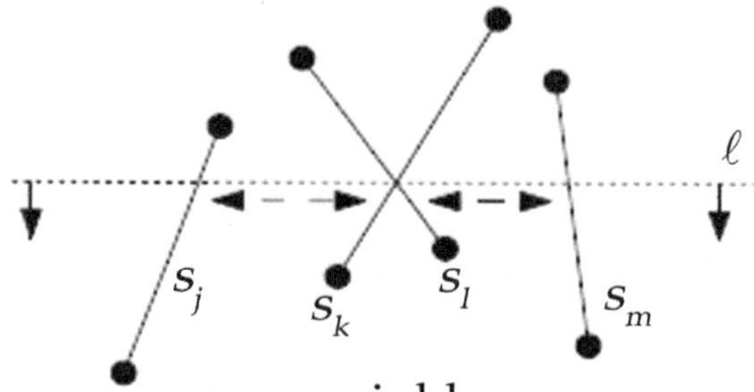

Figure 3.7. Intersection points and new neighbors.

Source: https://link.springer.com/book/10.1007/978-3-540-77974-2.

Allowing for the concept of being near in the horizontal path, let us rank the sections from left to right as they encounter the sweep line (Arge et al., 2008). We will only examine portions that are horizontally adjacent. This implies that every new segment is only compared to two existing sections, specifically those directly left and right of the higher terminus. Future on, whenever the sweep line descends, a segment may get next to additional segments that will be checked (Vogt and Bared, 1998). Our revised approach must be reflected in our algorithm's status, which should now correspond to the orderly arrangement of sections crossing the sweep line. The new status does not just apply to segment endings; it also applies to junction locations, where the sequence of crossed segments is altered. When this occurs, we must compare the two sections that have shifted positions to their unique neighbors. It is an entirely latest class of event point (Elliriki et al., 2019).

Prior to attempting to convert these concepts into an efficient method, we need to ascertain if the approach is accurate. We've decreased the amount of pairings to test, but are we still able to locate all intersections? Similarly, if two segments s_i and s_j overlap, is there constantly a sweep line location next to s_i and s_j laterally (Dévai, 1995)? Allow for some unpleasant cases first: presume that no section is horizontal, that every two segments interrelate at most once—they are not intersecting—, however no three segments interrelate at the same point. Later on, we will see that these instances are straightforward, but for now, it is best to disregard them. Whenever the sweep line approaches the ending of a segment, the crossings where the endpoint of one segment intersects another segment are readily spotted. Thus, the only remaining concern is whether crossings between segments' interiors are always identified (Fink and Wood, 1996).

- **Theorem 3.1:** Assume that s_i and s_j are two distinct segments whose inner sides cross at a single point p. So, there is an event position beyond 'p' at which s_i and s_j become contiguous as well as their intersection is checked (Yang and Yang, 1984).

Proof: Assume that is a horizontal line that is somewhat beyond p. If is sufficiently near to p, s_i, and s_j has to be contiguous laterally. (To become accurate, we might take like that there is no event spot on, among, or through p.) In different terms, there is a sweep line location at which s_i and s_j are contiguous (Brimkov et al., 2021). Instead, when the algorithm begins, s_i, and s_j are not yet adjacent, as the sweep line begins beyond all line segments as well as the status is empty. Thus, an event point 'q' must exist at which s_i and sj are becoming adjoining and are evaluated for crossroads (Figure 3.8) (Jokanovic, 2019).

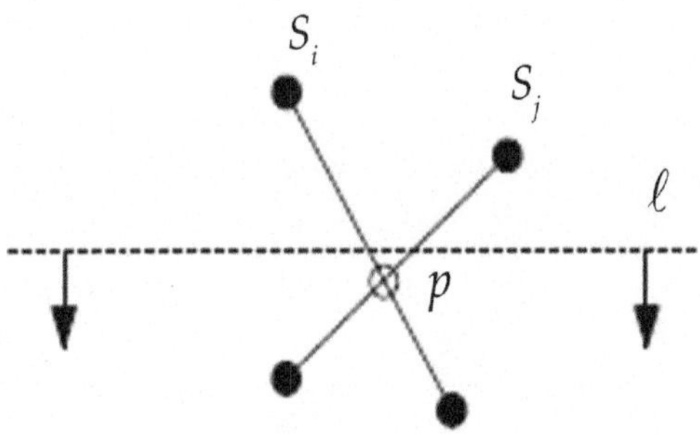

Figure 3.8. Two non-horizontal segments whose interiors intersect in a single point p.

Source: https://www.goodreads.com/book/show/316275.Computational_Geometry.

Thus, our method is accurate, at minimum when we disregard the heinous cases mentioned previously. Now, we can begin developing the plane sweep algorithm. Let us quickly recap the overall strategy. Consider sweeping a lateral sweep line downward across the plane (Devillers and Fabri, 1993). The sweep line comes to a halt at specified event locations; in our instance, these are the segment ends, that we understand in advance, and the junction sites, that are figured out on the fly. We retain the optimal route of segments crossed by the sweep line as it travels. When the sweep line comes to a halt at an event point, the arrangement of segments alters, as well as we must do multiple steps to update the status and identify junctions, based on the form of event point (Ashgriz and Poo, 1991).

Whenever the event point is a segment's higher endpoint, a new segment intersects the sweep line. This segment must be compared to its two neighbors all with sweep line for junction. Only the intersections underneath the sweep line are significant; those beyond the sweep line have already been found (Yi, 2012). If segments s_i and s_k are close on the sweep line and a new upper node of segment s_j arises among, we must examine s_j for overlap with s_i and s_k. We have discovered a new event point if we locate an interaction underneath the sweep line. After dealing with the higher endpoint, we go on to the next event point.

Whenever an event occurs at a junction, the two intersecting segments alter their order. Each of them receives (at most) one more neighbor against which intersection is evaluated (Skala, 1989). Once more, only crossings underneath the sweep line retain their interest. Assume that when the point of intersection s_k and s_l is reached, four segments s_j, s_k, s_l, and s_m seem to this order on the sweep line. Then s_k and s_l trade positions, and we must check for intersections between s_l and s_j, as well as s_k and s_m (Lu et al., 2005). Naturally, the latest crossings that we discover serve as event points for the algorithm. Though, it is conceivable that these occurrences have now been recognized, for example, if a set becomes together after previously being adjacent (Frankel et al., 2004).

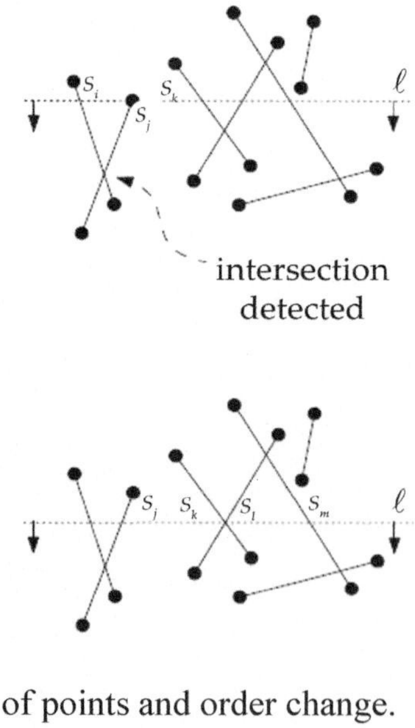

Figure 3.9. Intersection of points and order change.

Source: https://www.slideshare.net/BenjaminSach/line-segment-intersections.

Whenever the event point is the segment's lower terminal, its two adjacent neighbors are close and must be checked for intersection. If they cross underneath the sweep line, the point at which they cross is an event point. (Once again, this occurrence may have been spotted previously.). Let's suppose that when the lower terminus of s_l is met, three segments s_k, s_l, and s_m occur in this sequence on the sweep line (Figure 3.9) (Tilove, 1980; Devillers and Fabri, 1993).

Then s_k and s_m will become adjacent, and we will perform an intersection test on them. We have calculated all points of intersection once we have pushed the entire plane—or, more accurately, once we have handled the very last event point. This is ensured by the very next invariant, which remains true at any point at the time of plane sweep: all interconnection points beyond the sweep line have indeed been properly calculated (Kowaluk and Majewska, 2015).

Following this drawing of the algorithm, it's interval to delve deeper. Additionally, it is time to consider the depraved instances that can occur, such as three or even more segments intersecting at a point. In these instances, we must first clearly state how much we anticipate from the algorithm (Boissonnat and Snoeyink, 2000). We could need the algorithm to notify every intersection point just once, and it looks more beneficial if it claims a summary of segments that transfer along or use this as an endpoint for each intersection point. There is also another special condition in which we ought to be more precise in defining the desired outputs, namely two half overlapping segments, but for the sake of simplification, we will overlook this specific instance for the remainder of this section (de Luna et al., 2014).

To begin, we will describe the data structures that the algorithm utilizes. To begin, we require a data structure for storage events, termed " the event queue. The event queue is denoted by the letter Q. We require a procedure that extracts the very next event which will take place from 'Q' and goes back it for processing (Imai and Asano, 1984). This is the greatest event that occurs underneath the sweep line. If two event points possess identical y-coordinates, the one with the smaller x-coordinate is brought back. Similarly, event points that occur on similar horizontal line are handled in reverse order. This assumes that we must regard a horizontal segment's left endpoint as its upper node and its right ending point as its lower node. You could also consider our convention in this way: rather than a horizontal sweep line, assume it sloping slightly upward (Hoff III et al., 1999). Consequently, the sweep line intersects with the left end-point of a lateral segment just beforehand it intersects with the right node. Even though new events would be calculated on the fly, the event queue should indeed enable for insertions. Take note that two occurrences can occur concurrently. For instance, two separate segments' upper endpoints may overlap. It is more convenient to consider this as a single event point. As a result, an integration has to be capable of determining whether an event previously exists in 'Q' (Figure 3.10) (Samet and Webber, 1988).

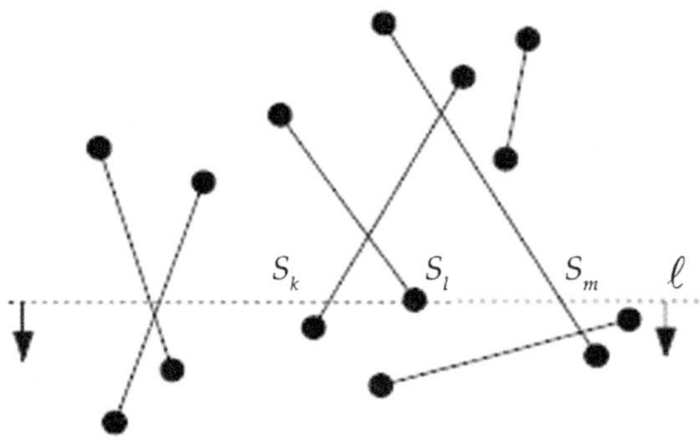

Figure 3.10. Whenever the lesser endpoint of s_l is encountered, three segments s_k, s_l, and s_m appear in order on the sweep line.

Source: https://www.slideshare.net/BenjaminSach/line-segment-intersections.

The event queue is implemented as follows. On the event points, define an order that corresponds to the sequence in which they're managed. Thus, if 'p' and 'q' are two event points, we possess p q only if $p_y > q_y$ stands true or $p_y = q_y$ holds true and p_x q_x holds true. The event points are stored in a balanced binary search tree in the order specified by. We would save the segments beginning at each event point 'p' in 'Q,' which really is, the segments for whom the upper node is p (Li and Dong, 2010). This data would be required to manage the event. Together retrieving the next event and trying to insert an event take O(logm) time, in which 'm' is the total quantity of events in Q. (We avoid using a heap to create the event queue since we need to be capable to determine whether a particular event already exists in Q) (Kelly, 1987).

Next, we must retain the algorithm's state. It is the series of segments that cross the sweep line in an orderly fashion. The status structure, represented by T, is used to gain reach to the neighbors of a particular segment 's' in order to perform intersection tests on them (Argel et al., 1995). The status structure has to be vibrant: when segments begin or end their intersection with the sweep line, they should be added to or removed from the structure. Since the status structure's segments have a well-defined order, we could use a self-balancing binary search tree as the status structure. This may come as a surprise to those accustomed to binary search trees which hold numbers. However, binary search trees could indeed contain any group of elements as far as the elements are ordered (Figure 3.11) (Ghosh and Mount, 1991).

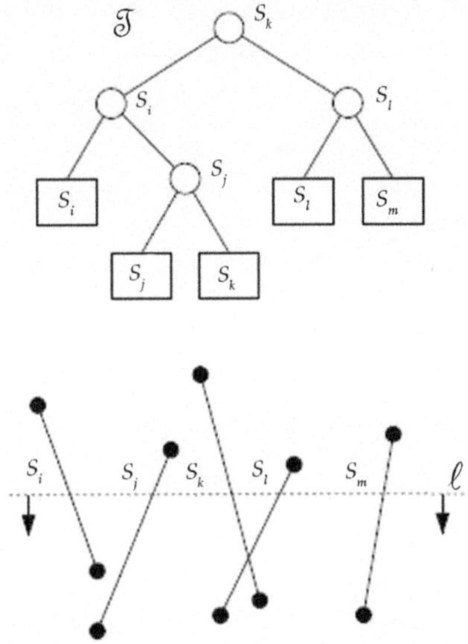

Figure 3.11. A balanced binary search tree with an orderly arrangement of segments interconnecting the sweep line and sections interconnecting the sweep line sorted in the leaves.

Source: https://www.springerprofessional.de/en/line-segment-intersection/2940436.

More precisely, we contain the sections interconnecting the sweep line in the leaves of a balanced binary search tree 'T' in the sequence in which they intersect. The sequence of segments all along the sweep line relates to the sequence of leaves in T.

Additionally, we must save data in the internal nodes to direct the search down the tree to the leaves. We contain the segment from a right-hand leaf in the left subtree of each internal node (Requicha and Voelcker, 1985).

(Put another way, we might contain the segments exclusively within the interior nodes.). It will free up a few storages space. Though, this is more intuitive to consider of the segments in inner endpoints as values that guide the search, rather than as data items. Additionally, keeping the segments in the leaves simplifies the description of some algorithms.) Assume we are looking for the segment instantly to the left of a point 'p' on the sweep line in T.

We determine whether 'p' is to the right or left of the segment saved at every inner node 'v' (Hoel and Samet, 1994). Dependent on the output, we decline to either the left or right subtree of and eventually reach a leaf. This leaf, or the leaf directly to its left, contains the segment we're looking for. Similarly, we can locate the segment stands to the right of p, as well as the segments that contain p.

As a result, every update and neighbor search operation are O(logn) in duration (Wegman, 1990).

We only require two data structures: the event queue 'Q' and the status structure 'T.' The global algorithm is now as given.

- **Algorithm Input FINDINTERSECTIONS(S). In the plane, a set 'S' of line segments.**

Output: The group of points of intersection between the segments in S, along with the segments that comprise every intersection point (Hertel and Mehlhorn, 1983).

- Create a new blank event queue Q. Following that, add the segment nodes into Q; whenever an upper node is implanted, the correlating segment must also be stored;
- Set a new status structure 'T' with an empty value;
- as long as 'Q' is not blank;
- do Delete the subsequent event point 'p' in 'Q;'
- HANDLEEVENTPOINT(p).

We've been seeing how events are dealt: at segment ends, we must add or remove segments through the status structure T, and at segment intersections, we must rearrange two segments.

In both cases, intersection tests among segments which become neighbors as a result of the event are also required.

The details become a little more complicated in debased cases—when multiple segments are engaged in a single event point (Barnhill et al., 1987). The following procedure explains how to properly handle event points; it is demonstrated in Figure 3.12.

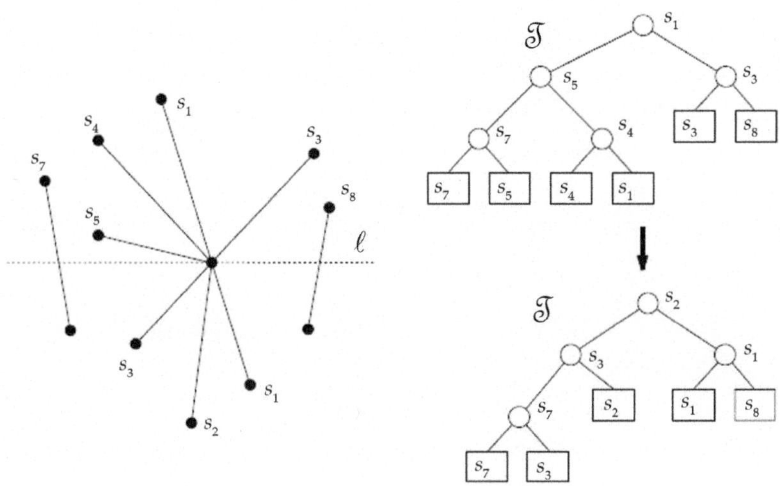

Figure 3.12. Illustrates an event and a resulting modification in the status structure.

Source: https://www.springerprofessional.de/en/line-segment-intersection/2940436.

- **HANDLEEVENTPOINT(p):**
- Define U(p) as the collection of segments with an upper endpoint equal to p; such segments are associated with the event point p. (In a case of parallel segments, the top end destination is always the left endpoint).
- Locate entirely segments in 'T' that comprise the character p; they must be adjacent in T. Let L(p) signify the subcategory of segments discovered with a lower node of p, and C(p) signify the subset of segments discovered with an interior containing p.
- If L(p) ∪U(p) ∪C(p) consists of multiple segments.
- Then along with L(p), U(p), and C, describe 'p' as an intersection (p).
- Remove the L(p) ∪C(p) segments from T.
- Add the U(p) ∪C(p) segments into T. The sequence of the segments in 'T' must correlate to the sequence in which a sweep line just under 'p' intersects them. If a horizontal segment exists, it is the last segment that contains 'p' (Ruzon and Tomasi, 2000).
- (Reversing the order of the segments of C(p) by deleting and re-inserting them).

- If U(p) ∪C(p) = 0, then.
- Then, let s_l and s_r be p's left and right neighbors in T, respectively.
- IDENTIFYNEWEVENT (s_l, s_r, $_p$).
- Else Assume that 's' is the left side segment of the U(p) ∪C(p) chain in T.
- Assume that s_l is s's left neighbor.
- FINDNEWEVENT(s_l, s', p)T.
- Assume that 's' is the right-hand segment of the U(p) ∪C(p) chain in T.
- Assume that s_r is s's right neighbor.
- FINDNEWEVENT(s'', s_r, p)

Take observation that lines 8–16 make the assumption that s_l and s_r occur. If they really do not occur, the associated stages should clearly be skipped (Deans, 1981).

The processes for discovering unique intersections are straightforward: they just conduct an intersection test between two segments. The only point we must be cautious of is whether a crossing has been addressed or not. Since there are no linear segments, when the point of intersection is below the sweep line, the intersection has still not been managed yet. However, how should we handle horizontal segments? Remembrance to our convention of treating events with similar y-coordinate in the left-to-right direction. It indicates that we are even now involved in points of intersection to the right side of the latest event point. As a result, the process FINDNEWEVENT is defined in the following manner (Pfoser and Jensen, 1999).

- **FINDNEWEVENT(s_l, s_r, p):**
- If s_l and s_r interrelate underneath the sweep line, or around it and to the right-hand side of the latest event point p, and the intersection has not yet been recorded in 'Q' as an event.
- Then, as an event, add the intersection point into Q.

What about our algorithm's correctness? While it seems obvious that FINDINTERSECTIONS returns just true intersection points, does it return all together to them? The following theorem establishes that this is actually true (Crowley, 1985).

- **Theorem 3.2:** The algorithm FINDINTERSECTIONS correctly calculates all points of intersection and their associated segments.

Proof: Recall that an event's priority is determined by its y-coordinate, and whenever two events have had the similar y-coordinate, the one with the smaller x-coordinate is prioritized. We will demonstrate the theorem by induction based on the order of the event points (Xu and Zhang, 1996).

Presume that almost all interconnection points 'q' with such a top importance have been correctly calculated. We will demonstrate that 'p' and the segments containing 'p' are properly calculated. Let U(p) denote the set of segments with 'p' as their top end-point (or, in the case of parallel segments, their left node), L(p) denote the group of segments with 'p' as their lesser end-point (or, in the case of parallel segments, their right-side endpoint), and C(p) denote the group of segments with 'p' in their inner (Hillier and Iida, 2005).

To begin, let us presume that 'p' is the end-point with one or even more segments. In that case, at the beginning of the algorithm, 'p' is saved in the event queue Q. The segments from U(p) are saved with 'p' to ensure their discovery. When 'p' is handled, the sections from L(p) and C(p) are saved in T, and thus appear in line 2 of HANDLEEVENTPOINT. Thus, whenever 'p' is an end-point of one or both of the segments, p, and then all the sections engaged are correctly determined (Brown, 1979).

Suppose that 'p' is not a segment endpoint. All that is required is to demonstrate that 'p' would be entered into 'Q' at a certain point. Take note that each section that is involved contains a 'p' in its core. Arrange these segments according to their angle about p, and consider s_i and s_j to be two adjoining segments. As a result of the demonstration of Theorem 3.1, we observe that there is a greater importance event point as compared to 'p' like s_i and s_j remain contiguous whenever 'q' is passed. For the sake of simplicity, we assumed the si and sj are not horizontal segments in Theorem 3.1, although the proof is easily adaptable to horizontal segments. The event point 'q' was appropriately managed by induction, that indicates that 'p' was noticed and saved in 'Q' (Berg et al., 1997).

As a result, we now have a proper algorithm. However, did we manage in constructing an algorithm that is output-sensitive? Yes, the method runs in O((n+k)logn) time, wherein 'k' is the output size. The following theorem establishes an even stronger conclusion: the running time is O((n + I)logn), wherein 'I' is the intersection count. This is more robust, because the output for a single intersection point might contains a lot of segments, as is the case when many segments collide at a common location (Sarnak and Tarjan, 1986).

- **Theorem 3.3:** Algorithm FINDINTERSECTIONS has a running time of O(nlogn+Ilogn) for a set 'S' of 'n' line segments in the plane, where 'I' is the number of intersection sites between segments in 'S' (Aggarwal et al., 1988).

Proof: The algorithm begins by populating the event queue with events from the segment endpoints. This takes O(nlogn) time since we constructed the event queue as a balancing binary search tree. Starting the status structure consumes an indefinite amount of time. The plane sweep then begins, and all further occurrences are addressed. To process an event, we conduct three actions on the event queue Q: the event is removed from 'Q' in line 4 of FINDINTERSECTIONS, and one or two calls to FINDNEWEVENT may result in the addition of up to two new events to 'Q' (Oyler et al., 2014). Each deletion and insertion operation on 'Q' takes O(logn) time. Additionally, we conduct operations on the status structure T—insertions, deletions, and neighbor discovery—that each require O(logn) time. The quantity of operations is proportional in the quantity of segments participating in the event, m(p):= card(L(p)U(p)C(p)). If we use the symbol 'm' to signify the sum of all m(p) across all event points p, the algorithm's execution time is O. (mlogn) (Barnard, 1983).

It is self-evident that m = O(n+k), where 'k' is the output size; after all, anytime m(p) > 1, we send all segments included in the event, but the only events including a single segment are segment nodes. However, we wish to demonstrate that 'm' equals O(n+I), wherein 'I' is the amount of intersecting points. To demonstrate this, consider the collection of segments as a planar graph contained in the plane (Figure 3.13) (Lane and Riesenfeld, 1980).

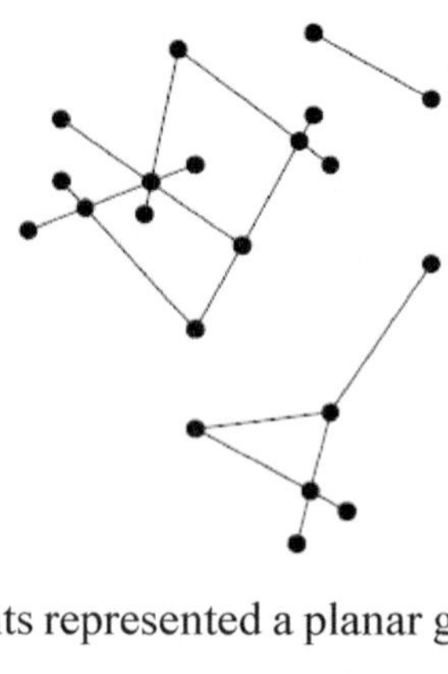

Figure 3.13. Set of segments represented a planar graph embedded in the plane.

Source: https://www.sciencedirect.com/topics/computer-science/planar-graph.

Its own vertices are segment endings and junction sites, and its sides are segment parts that link the vertices. Consider the pth event point. This is a vertex of the graph, and m(p) is constrained by the vertex's degree. As a result, 'm' is constrained by the total of the levels of all of the vertices in our network (Miller, 2005). Because each graph edge adds one to the grade of precisely two vertices (its nodes), 'm' is constrained by 2ne, wherein ne is the set of corners in the graph. Let us constrain ne in respect of 'n' and I. The amount of vertices, nv, is defined to be no more than 2n + I. It is commonly known that ne = O(nv) in planar graphs, proving our assertion (De Berg et al., 2000). But, for the sake of wholeness, assume provide the argument here. Each facet of the planar graph is bordered by at minimum three edges (assuming there are at minimum three segments), and an edge can only bind two separate faces. As a result, the amount of facets, nf, is no more than 2ne/3. Now we use Euler's formula, which stipulates that for every planar network having n_v vertices, n_e edges, and n_f faces, the (Lasser, 1986):

$$2 \leqslant (2n+I) - n_e + \frac{2n_e}{3} = (2n+I) - n_e/3$$

So n_e 6n + 3I − 6, and 'm' 12n + 6I − 12, and the bound on the running time follows.

We still need to investigate the second part of difficulty, the storing space needed by the method. Because the tree 'T' only saves a segment once, it employs O. (n) memory. However, the extent of 'Q' can be increased. The algorithm adds the intersection given relation (Chen and Pottmann, 1999):

$$n_v - n_e + n_f \geq 2$$

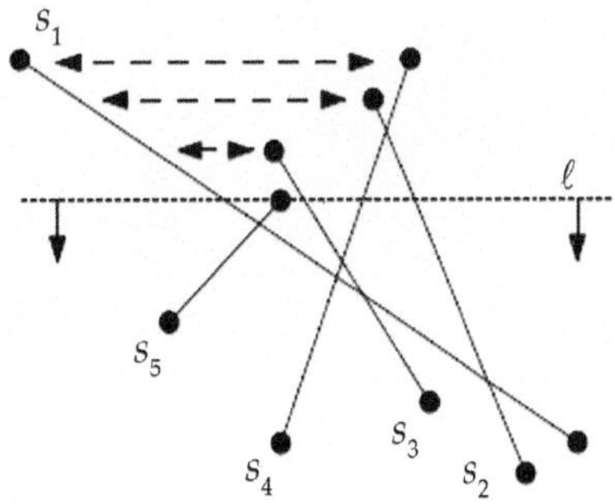

Figure 3.14. Intersection positions of sets of segments on the sweep line that are presently adjacent.

Source: https://slideplayer.com/slide/16270878/.

However, if the graph is linked, equality stands Whenever the bounds on nv and nf are entered into this formula, we obtain points in 'Q' whenever they are recognized and points are removed whenever they are managed (Güting and Schneider, 1993). When crossings take a long time to resolve, 'Q' can balloon to enormous proportions. Of course, its size has always been constrained by O(n+I), but it'd be preferable if the active storage was always linear (Figure 3.14) (Schoen, 1970).

3.3. THE DOUBLY-CONNECTED EDGE LIST

The simplest claim of the map overlay issue has been fixed, in which the two maps are systems defined as items of line segments. Maps, in overall, have a more compound structures: they are labeled sectors of the plane. A theme of Canadian forests, for example, would be a partition of the country into areas with tags like "pine," "deciduous," "birch," and "mixed" (de Berg et al., 2008).

We must first create a good recognition for a subdivision until we can offer an algorithm for calculating the overlay of two sectors. It's not a smart idea to save a segmentation as a selection of line segments. Operations like reporting a region's boundary would be extremely difficult. It's preferable to include structural and topological data, such as which segments bind a given group, which areas are adjoining, and so on (Figure 3.15) (Keeler et al., 2007).

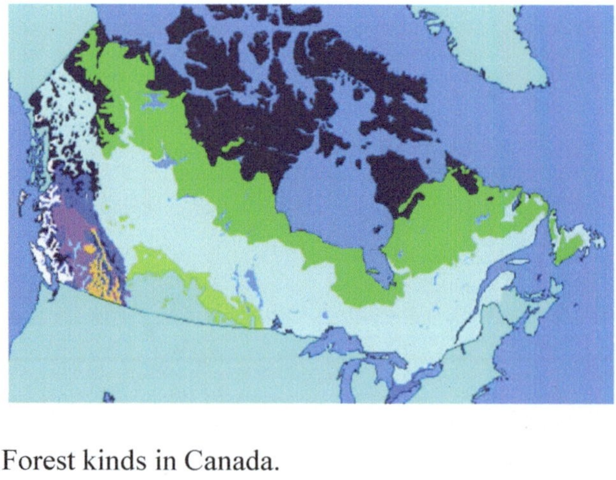

Figure 3.15. Forest kinds in Canada.

Source: https://www.nrcan.gc.ca/our-natural-resources/forests/sustainable-forest-management/measuring-and-reporting/forest-classification/13179.

The maps we're looking at are planar subdivisions of graphs caused by planar embeddings. If the implicit graph is linked, this same subdivision is linked. A vertex is the incorporation of an endpoint in the graph, and an edge is the incorporation of an arc (Razzazi and Fotouhi, 2010). We only take into account embeddings in which each edge is a line segment. In theory, the edges of a sector do not have to be horizontal.

A partition does not have to be a planar graph incorporation because it can have unlimited edges. Nevertheless, such more generic groupings are not considered in this part. An edge is considered open if its nodes are subdivision vertices—do not form part of it. A maximum linked subgroup of the plane that does not include a point on an edge or a vertex is called a face of the subdivision (Wong and Müller, 1991).

As a result, a facing is an open polygonal area whose boundary is defined by the subdivision's edges and vertices. The amount of vertices, edges, and faces of a subdivision are added together to determine its complexity. When a vertex is the terminus of an edge, the vertex and the edge are said to be incident. A facial as well as an edge on its boundary are also incident, as are a face and the vertex of its border (Figure 3.16) (Preparata and Shamos, 2012).

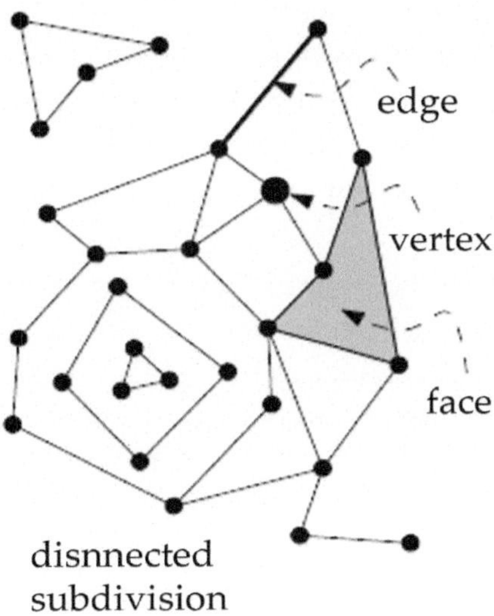

Figure 3.16. Planar subdivisions of graphs caused by planar extracted feature.

Source: https://slideplayer.com/slide/8132261/.

What should we expect from a divisional representation? Evaluate the face comprising a specific point is one operation that could be requested. This is obviously helpful in some implementations, we'll layout a data structure for it in a later chapter—but it's a little too much to expect from a basic illustration. Things we may request need to be more localized (Berg et al., 1997). For instance, it is logical to expect that we be capable of walking around the perimeter of a given facet or that we be able to obtain one face from another if we are provided a familiar edge. Some other helpful operation is to consult all the edges surrounding a given vertex. These operations are supported by the illustration that we will discuss. The doubly-connected edge description is what it's called (Halperin and Sharir, 2017).

So, every face, edge, and vertex of the sub-division is represented by a record in a doubly-connected edge list. Aside from the geometric as well as topological data that will be discussed later, every document might also save extra data. If the subdivision symbolizes a vegetation thematic map, the doubly connected edge description would save the kind of vegetation in every face file for the respective region (Devai, 1986; Mittal and Bhowmick, 2012). The additional data is also referred to as attribute data.

We should be likely to function the basic operations pointed out earlier using the geometric as well as topological data held in the doubly connected edge list. We save a pointer from every edge to another to be capable of walking around with a face in anticlockwise sequence. We also keep a pointer to the preceding edge in case we need to walk together around face some other reason. Because an edge typically connects two faces, we'll require two pointers for it (Vassilev and Eades, 2013).

It's easier to think of the various viewpoints of an edge as two separate half-edges, so that each half-edge has its own upcoming half-edge and preceding half-edge. Also, it implies that a half-edge can only bind one face at a time. Twins are the two half-edges we do for a provided edge (Fan et al., 2011; Hanniel and Halperin, 2000). A next half-edge of a given half-edge is defined with regard to an anticlockwise traversal of a face, resulting in an alignment on every half-edge: it is directed to the face it confines is to the left of a viewer walking all along the edge. Even though half-edges are oriented, we can talk about their origin and the destination. If a half-origin edge's and destination are 'v' and 'w,' then it's a dual Twin's origin and destination are 'w' and 'v.' We only have to save one pointer in the face file to a random half-edge enclosing the face to attain its boundary. Starting at that half-edge, we can walk all around face by stepping from one half-edge to the next (Dobrindt et al., 1993). The opposite is true for limits of gaps in a face: if they are transited in anticlockwise order, the facet remains to the right. We alter the course of traversal for the border of an opening to clockwise because it will be easier to arrange half-edges so that their facet constantly remains to the similar side. Now, any half-edge on a face's boundary is always with the left of it (Kim and Shan, 2011). Some other result is that dual half-edges are always oriented in opposite directions. Given the existence of openings in a face, one pointer from of the facet to a random half-edge on its limit is insufficient to check the entire boundary: each boundary component requires a pointer to a half-edge. We can also save pointers to isolated vertices that don't have any incident edges if a face has them. This case will be ignored for the sake of simplicity (Fogaras and Sugihara, 2002).

Let's sum it up. There are three collections of records in the doubly-connected edge list: one for vertices, one for faces, and one for half-edges. The very next geometric and topological data is stored in this record keeping:

- The coordinates of a vertex 'v' are stored in a field called vertex record (Lazarus et al., 2001). Coordinates(v). It also keeps a pointer to a random half-edge with 'v' as its origin, IncidentEdge(v).
- A pointer OuterComponent(f) to a certain half-edge on a face's outside border is stored in its face record. This pointer is nil for the unbounded face. It also keeps a collection InnerComponents(f), that includes a pointer to a half-edge on the hole's border for each hollow on the face (Alt et al., 2005).
- A pointer Origin to a half-origin, edge's a pointer Twin to its twin half-edge, and a pointer IncidentFace to the face which, it limits are all stored in the half-edge record. We don't require to keep track of an edge's destination since it's the same as Origin(Twin) (Ilinkin et al., 2001). When traversing from origin to destination, the source is chosen so that IncidentFace is to the left of e. The half-edge record also contains references to the next and preceding edges on IncidentFace's boundary, respectively. Thus, Next is the only half-edge on the IncidentFace boundary which must contain endpoint of 'e' as its source, while P_{rev} is the only half-edge on the IncidentFace boundary that has Origin as its destination (Fogel et al., 2004).

Every vertex and edge receive the same amount of data. Because the list InnerComponents(f) includes as many members as the number of openings in the face, a facet may need more space (Sherbrooke et al., 1996). We find that the quantity of space is linear in the difficulty of the subdivision since each half-edge is directed to about once from across all InnerComponents(f) lists combined. The following is an instance of a doubly connected edge listing for a basic subdivision. Incident Face's two half-edges relating to an edge e_i are labeled as e_i,1 and e_i,2 (Goodrich, 1992).

To accomplish the fundamental tasks, the data contained in the doubly-connected edge collection is sufficient. For instance, beginning at the half-edge Outer Component, we can traverse around the outside border of a face image by coming Next pointers (f). All edges adjacent to a vertex 'v' can also be visited. It's an excellent exercise to find out how to accomplish this on your own (Driscoll, 1996).

3.4. COMPUTING THE OVERLAY OF TWO SUBDIVISIONS

We could now start tackling the overall map overlay issue now that we've created a fair description of a subdivision. The overlap of two sectors S_1 and S_2 is described as the subdivision $O(S_1,S_2)$ with a face 'f' an if only when there are facets f_1 in S_1 and f_2 in S_2 so that 'f' is a maximum possible linked subset of f_1 f_2. The overlap is the sector of the plane caused by the corners from S_1 and S_2 (Guibas and Seidel, 1987). This makes it sound more complex than it really is: just what signifies would be that the overlay is the sector of the plane caused by the ends from S_1 and S_2. This is depicted in Figure 3.17.

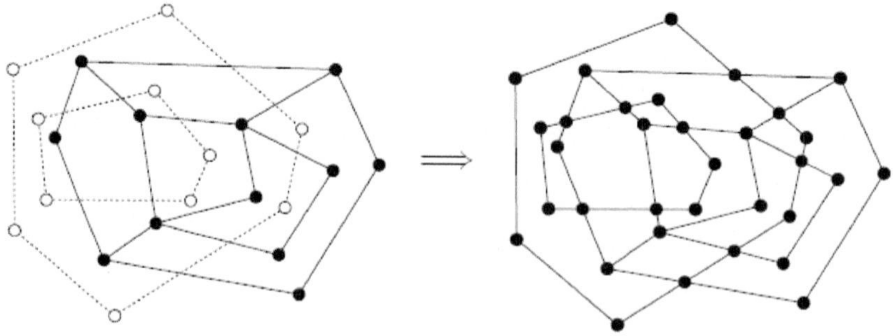

Figure 3.17. Overlaying two subdivisions.

Source: https://www.powershow.com/view/2237e-Y2NlM/Overlay_of_Two_Subdivisions_powerpoint_ppt_presentation.

Provided the doubly connected edge entries of S_1 and S_2, the overall map overlap issue is to calculate a doubly-connected edge collection in $O(S_1,S_2)$. Every face in $O(S_1,S_2)$ must be labeled with the tags of the face images in S_1 and S_2 that it contains. As a result, we have reach to the attribute data for these facets. It would imply that we understand the kind of vegetation and the number of precipitations for each region in an overlap of a vegetation map and a rainfall map (Fabri and Pion, 2009).

And see how much data we can recycle from either the doubly-connected edge entries for S_1 and S_2 in the doubly-connected edge record for 'O' first (S_1,S_2) (Berg et al., 1997). Take a look at S_1's system of edges and vertices. The edges of S_2 reduced this net into parts. These parts can be reused for the most part; only the edges reduce by the edges of S_2 should be replaced. Is this true, however, for the half-edge data that correlate to the pieces in the doubly-connected edge list? We'd have to variate information in these records if the direction of a half-edge changed (Samet and Webber, 1988). Thankfully, this isn't the case. The half-edges are focused so that the bound face is on the left; the face's shape may alter in the overlay, but that it will always be on the similar side of the half-edge. As a result, half-edge records for edges that are not overlapped by edges from further map can be reused. To put it another way, the only half-edge documents in the doubly-connected edge ranking for $O(S_1,S_2)$ that we can't borrow from S_1 or S_2 are those which occur when edges from various maps intersect (Güting, 1994).

The crossings in a group of line segments are calculated using the plane sweep algorithm described in Section 3. The sequence of segments which is the addition of pairs of edges of the two subdivisions S_1 and S_2 is used to run this algorithm (Zhang and Ghosh, 2000). We suggest the edges to be shut in this case. Remember that the algorithm is reinforced by two data structures: an event queue Q, that also saves the event points, and a balanced binary search tree T, that also stores the segments interconnecting the sweep line, decided to order from left to right. A doubly-connected edge listing 'D' is now also maintained. 'D' starts out with a copy of S_1's doubly-connected edge list and a copy of S_2's doubly-connected edge ranking. We'll convert 'D' into an accurate doubly-connected edge ranking for 'O' at the time of the plane sweep (S_1,S_2) (Lane and Riesenfeld, 1980). That is, the vertex and half-edge records will be calculated first, and the face data will be calculated later. Among the edges of the status structure 'T' and the half-edge data in 'D' that correlate to them, we keep cross pointers. When we come to an intersection point, we can get to the part of 'D' that requires to be reformed this way. The invariant we sustain is that the portion of the overlay just beyond sweep line has been calculated properly at all times during the sweep (Dueker, 1987).

Let's think about what we should do whenever we attain an event point. To begin, we upgrade 'T' and 'Q' in the same way that the line segment crossing algorithm does. It is all; the event point is a re-usable vertex if the event single includes limits from one to two sectors. We have to allow local variations to 'D' to connect the doubly-connected edge records of the two original subdivisions at the point of intersection if the enhancing edges from both subdivisions. It's time-consuming but not difficult (Park and Shin, 2002).

Figure 3.18 shows the details of one of the suspected options, where an edge 'e' of S_1 goes through a vertex 'v' of S_2. Two edges, represented 'e'' and 'e'',' must be substituted for the edge 'e.' The two half-edges for 'e' need to make four in the doubly-connected edge listing. We make two new half-edge files, one with 'v' as the source and the other without (Van Den Bergen, 2003). As seen in Figure 3.18, the two known half-edges for 'e' maintain the nodes of 'e' as their source. Then, by setting their Twin() pointers, we couple up the current half-edges along the latest half-edges. As a result, 'e' is defined by one unique and one current half-edge, and vice versa. We'll need to set a few P_{rev}() and Next() pointers now (Held, 2001). We'll start with the condition surrounding e's endpoints and then move on to the situation surrounding v. Each of the two latest half-edges' Next() pointers copies the Next() pointer of the old half-edge which is not its identical. These half-edges' P_{rev}() pointers have to also be updated and set to the new half-edges. The easiest way to check if this step is accurate is to look at a diagram (Requicha and Voelcker, 1985).

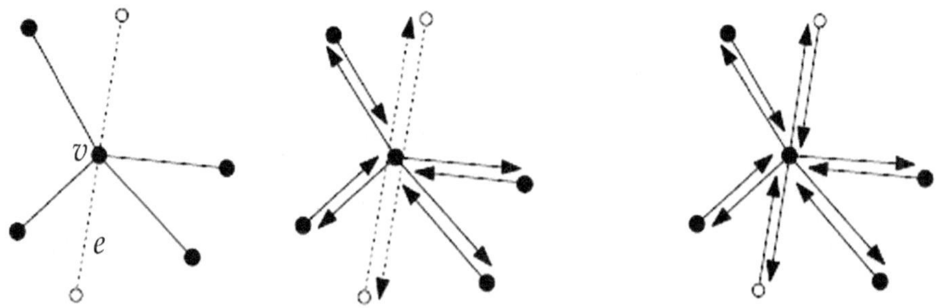

Figure 3.18. An edge of one sector passing through a vertex of another.

Source: https://www.cl.cam.ac.uk/teaching/2004/AdvGraph/exercise2.html.

It is still necessary to fix the condition surrounding vertex v. The pointers Next() and Prev() of the four half-edges chosen to show 'e' and 'e,' as well as the four half-edges occurrence from S_2 to v, must be set. We test in which 'e' and 'e' must be in the serial order of the corners all over vertex 'v' to find such four half-edges from S_2 (Lane and Riesenfeld, 1980). A Next() pointer from one half-edge as well as a Prev() pointer from other connect the four half-edges together. Suggest the 'e' half-edge with 'v' as its final destination. It has to be connected to the first half-edge, been seeing circular from e, and with 'v' as its source. The first anticlockwise half-edge with 'v' as its destination must be connected to the half-edge for 'e' with 'v' as its origin. The similar can be said about 'e' (Dueker, 1987).

The majority of the steps in the preceding explanation take the same amount of time every time. It may take longer to locate in which 'e' and 'e' seem so in the serial order about v: it will have some time-sequential in the degree of v. Intersections of two corners from various maps, and happened to coincide vertices, for example, are not any more challenging as compared to the particular instance we just mentioned (Held, 2001). These situations also have an O(m) time complexity, wherein 'm' is the total of edges that intersect the event point. As a result, upgrading 'D' does not enhance the line segment intersection algorithm's running time exponential. Every interconnection we come across is a set of vertices of the overlay. As a result, the vertex records and half-edge files of the doubly-connected edge ranking for $O(S_1,S_2)$ can be calculated in O(nlogn+klogn) time, where 'n' signifies the sum of S_1 and S_2 complexities, and 'k' signifies the overlay complexities (Requicha and Voelcker, 1985).

That after fields regarding vertex and half-edge data have indeed been group, the data regarding O's faces must be computed (S_1,S_2). To be more specific, we must build a face record in $O(S_1,S_2)$ for every face f, make OuterComponent(f) point to a half-edge on the outside boundary of 'f,' and establish a collection InnerComponents(f) of pointers to half-edges on the borders of the holes within f. In addition, the IncidentFace() elements of the half-edges on the border of 'f' must be set to refer to the face data of 'f.' Lastly, every one of the fresh faces have to always be labeled along the identities of the faces which it replaces in the old sectors (Held, 1997; Auger and Lawrence, 1989). We create a graph 'G' to see which boundary cycles are tied to the same face. There is an endpoint in 'G' for each boundary cycle, both inlet and outlet. There is also an endpoint for the unbounded face's imaginary outer boundary. When one cycle is the boundary of a gap and the further cycle has a half-edge instantly to the left of the upper left vertex of that whole cycle, there is an arc among them (Mairson and Stolfi, 1988). Since there is no half-edge to the left of the cycle's upper left vertex, the cycle's node is connected to the unbounded face's node. It described in Figure 3.19. The patterned segments in the diagram show how the hole cycles are linked to other cycles. The figure also includes a graph that corresponds to the subdivision. The inner boundary cycles are represented by on its own circles, while the external boundary cycles are represented by double circles. C_3 and C_6 are in the similar linked element as C_2 as you can see (Arge et al., 1995; Fogel et al., 2004). This means C_3 and C_6 are hole cycles in the face with C_2 as the external surface. If a face 'f' has only one hole, the graph 'G' connects the hole's boundary cycle to the face's external surface. In summary, this is not the case: as shown in Figure 3.19, a hole can be connected to some other hole. This hole, which would be located on the same face as 'f,' could be connected to f's outer boundary or go to another hole. However, as the next Theorem 3.4 demonstrates, we must ultimately link a hole to the external surface (Güting and Schneider, 1995).

- **Theorem 3.4:** The pair of cycles event to one face basically relates to every node of the graph G.

Proof: Suggest the boundary of a cycle 'C' crossing an opening in a face 'f.' 'C' have to be connected to some other cycle that also restricts 'f' even though 'f' tells a lie directly to the left of the leftmost vertex of C. As a result, the same face was bound by cycles in the similar linked element of 'G' (Hoel and Samet, 1992).

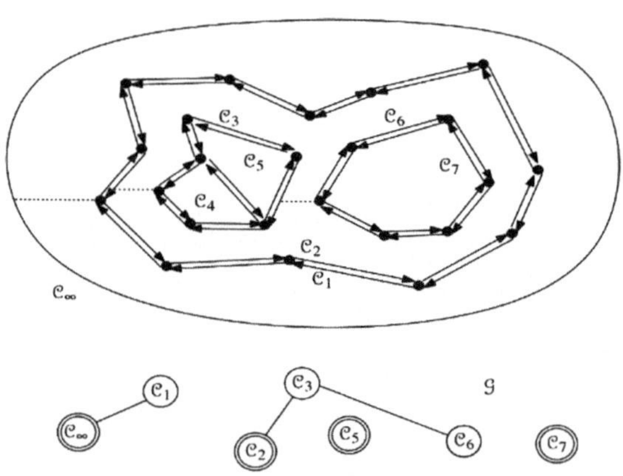

Figure 3.19. A subdivision and the corresponding graph G.

Source: http://www.cs.uu.nl/geobook/toc.html.

We prove that each cycle bordering a gap in 'f' is in the similar linked part as f's external boundary to complete the proof. Let's pretend there's a cycle when this isn't the case. Suppose 'C' be the leftmost like that cycle, with the leftmost vertex being the leftmost. There is, by necessity, an arc among 'C' as well as another cycle 'C' which is somewhat to the left of C's leftmost vertex. As a result, 'C' is in the same linked element as C, that is not the outer border element of f. This is in direct contradiction to the definition of 'C' (De Floriani et al., 2000; Newell et al., 1972).

When all of the pieces are put collected, we obtain the next algorithm.

- **MAPOVERLAY is a mapping algorithm (S_1,S_2)**

Input: In doubly-connected edge lists, two plane subdivisions S_1 and S_2 are kept.

Output: In a doubly-connected edge list D, the overlap of S_1 and S_2 is kept (White and Cohen, 1980).

- Duplicate the S_1 and S_2 doubly-connected edge categories to a unique 'D' doubly-connected edge list.
- Using the plane sweep approach from Section 3.2, calculate all crossings among edges of S_1 and S_2. Do the given in addition to the activities on 'T' and 'Q' that are needed at the event points:
    - If the event impacts all S_1 and S_2 edges, update 'D' as described beyond. (In the scenario whenever an edge of S_1 crosses through a vertex of S_2, this was addressed).
    - Save the half-edge to the left of the event at the vertex in 'D' that represents it (Quirk, 1994; Baillard et al., 1999).
- (Nowadays 'D' is the doubly-connected edge collection for O(S_1,S_2), but where the face data has not yet been calculated∗).
- Traverse 'D' to find the boundary cycles in O(S_1,S_2).
- Calculate the linked parts of the graph G, where endpoints with respect to edge cycles and such whose arcs link every hole cycle to the left of its leftmost vertex. (The data needed to calculate G's arcs was calculated in line 2, second item).
- in G, for every linked section (Laidlaw et al., 1986).
- Let 'C' be the component's distinct outer boundary cycle, and 'f' signify the face bordered by the cycle. Make a face data for f, assign OuterComponent(f) to one of C's half-edges, then build a list InnerComponents(f) with pointers to one half-edge in every hole cycle in the element. Allow all half-edges in the cycles'

IncidentFace() pointers to point to f's face record.

- As previously mentioned, mark every face of $O(S_1,S_2)$ with titles of the surfaces of S_1 and S_2 that it contains (Chiang and Tamassia, 1992).
- **Theorem 3.5:** Let S_1 be a plane sector of difficulty n_1, S_2 be a sector of difficulty n_2, and $n := n_1 + n_2$ be a subdivision of complexity n_2. The overlap of S_1 and S_2 may be built-in O(nlogn+klogn) time, wherein 'k' denotes the overlap's difficulty (Theobald, 2001; Li et al., 1992).

Proof. By Theorem 2.3, duplicating the doubly-connected edge listings in line 1 requires O(n) time, while line 2's plane sweep requires O(nlogn+klogn) time. Steps 4–7, wherein we refill in the face data, take time linear in the difficulty of O(nlogn+klogn) (S_1,S_2). (A basic depth-first search may find the linked elements of a network in linear time.) Lastly, every face in the resultant subdivision may be labeled with the face of the initial sectors that include it in O(nlogn+klogn) time (Wu et al., 2000).

3.5. BOOLEAN OPERATIONS

The map overlap algorithm is a versatile tool that could be applied to a variety of other tasks. One in specific is necessary for making Boolean operations on two polygons P_1 and P_2. As an illustration, see Figure 3.20. Observe that the operations' output may no more be a polygon. It can be composed of a variety of polygonal areas, few of which contain holes (Requicha and Voelcker, 1985).

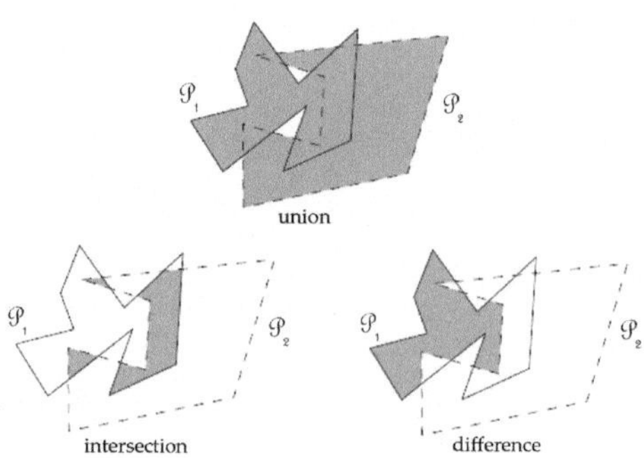

Figure 3.20. Union, difference, and intersection of Boolean operations on two polygons P_1 and P_2.

Source: http://www.cs.uu.nl/geobook/toc.html.

To operate the Boolean operation, we consider the polygons to be planar maps, with their bounded faces labeled P_1 and P_2. We create an overlap of such maps and then obtain the faces in the overlay for whom the labels correlate to the Boolean operation we're executing. To calculate the interconnection P_1P_2, we obtain the faces labeled with P_1 and P_2 from the overlay (Fabri and Pion, 2009). To calculate the union P_1P_2, we obtain the faces labeled to P_1 or P_2 from the overlay. And to calculate the differential P_1P_2, we obtain the face images in the overlap which are labeled along P_1 but not P_2.

Since each point at which a corner of P_1 intersects with a corner of P_2 is a vertex of P_1P_2, the algorithm runs in O(nlogn+klogn), in which 'n' is the overall set of vertices in P_1 and P_2, and 'k' is the difficulty of P_1P_2 (Rivero and Feito, 2000). The similar is true for all other Boolean operations: any connection between two corners is a vertex of the end outcome, regardless of the operation. As a result, we instantly obtain the given result.

Corollary Allow P_1 to be a polygon with n_1 nodes and P_2 to be a polygon with n_2 vertices, with $n = n_1 + n_2$. Then P_1P_2, P_1P_2, and P_1P_2 could be calculated in O(nlogn+klogn) time, in which 'k' denotes the output's difficulty (Gardan and Perrin, 1996).

REFERENCES

1. Aggarwal, A., Chazelle, B., Guibas, L., Ó'Dúnlaing, C., & Yap, C., (1988). Parallel computational geometry. Algorithmica, 3(1), 293–327.
2. Alt, H., Cheong, O., & Vigneron, A., (2005). The Voronoi diagram of curved objects. Discrete & Computational Geometry, 34(3), 439–453.
3. Arge, L., Mølhave, T., & Zeh, N., (2008). Cache-oblivious red-blue line segment intersection. In: European Symposium on Algorithms (Vol. 2, pp. 88–99). Springer, Berlin, Heidelberg.
4. Arge, L., Vengroff, D. E., & Vitter, J. S., (1995). External-memory algorithms for processing line segments in geographic information systems. In: European Symposium on Algorithms (3rd edn., Vol. 1, 2, pp. 280–320). Springer, Berlin, Heidelberg.
5. Argel, L., Vengroff, D. E., & Vitter, J. S., (1995). External-memory algorithms for processing line segments. In: Algorithms-ESA'95: Third Annual European Symposium, Corfu, Greece, Proceedings (Vol. 3, p. 295). Springer Science & Business Media.
6. Ashgriz, N., & Poo, J. Y., (1991). FLAIR: Flux line-segment model for advection and interface reconstruction. Journal of Computational Physics, 93(2), 449–468.
7. Auger, I. E., & Lawrence, C. E., (1989). Algorithms for the optimal identification of segment neighborhoods. Bulletin of Mathematical Biology, 51(1), 39–54.
8. Baillard, C., Schmid, C., Zisserman, A., & Fitzgibbon, A., (1999). Automatic line matching and 3D reconstruction of buildings from multiple views. In: ISPRS Conference on Automatic Extraction of GIS Objects from Digital Imagery (Vol. 32, pp. 69–80).
9. Barnard, S. T., (1983). Interpreting perspective images. Artificial Intelligence, 21(4), 435–462.
10. Barnhill, R. E., Farin, G., Jordan, M., & Piper, B. R., (1987). Surface/surface intersection. Computer-Aided Geometric Design, 4(1, 2), 3–16.
11. Bartuschka, U., Mehlhorn, K., & Näher, S., (1997). A robust and efficient implementation of a sweep line algorithm for the straight line segment intersection problem. In: Proc. Workshop on Algorithm Engineering (Vol. 1, pp. 1–15).
12. Berg, M. D., Kreveld, M. V., Oostrum, R. V., & Overmars, M., (1997). Simple traversal of a subdivision without extra storage. International Journal of Geographical Information Science, 11(4), 359–373.
13. Berg, M. D., Kreveld, M. V., Overmars, M., & Schwarzkopf, O., (1997).

Computational geometry. In: Computational Geometry (Vol. 1,4, pp. 1–20). Springer, Berlin, Heidelberg.

14. Boissonnat, J. D., & Snoeyink, J., (2000). Efficient algorithms for line and curve segment intersection using restricted predicates. Computational Geometry, 16(1), 35–52.
15. Bonet, J., & Peraire, J., (1991). An alternating digital tree (ADT) algorithm for 3D geometric searching and intersection problems. International Journal for Numerical Methods in Engineering, 31(1), 1–17.
16. Brimkov, B., Geneson, J., Jensen, A., Broussard, J., & Nowbandegani, P. S., (2021). Intersections and circuits in sets of line segments. Journal of Combinatorial Optimization, 2(3), 1–22.
17. Brown, K. Q., (1979). Voronoi diagrams from convex hulls. Information Processing Letters, 9(5), 223–228.
18. Chazelle, B., & Edelsbrunner, H., (1992). An optimal algorithm for intersecting line segments in the plane. Journal of the ACM (JACM), 39(1), 1–54.
19. Chazelle, B., Edelsbrunner, H., Guibas, L. J., & Sharir, M., (1994). Algorithms for bichromatic line-segment problems and polyhedral terrains. Algorithmica, 10, 11(2), 105–135.
20. Chen, H. Y., & Pottmann, H., (1999). Approximation by ruled surfaces. Journal of Computational and Applied Mathematics, 102(1), 143–156.
21. Chiang, Y. J., & Tamassia, R., (1992). Dynamic algorithms in computational geometry. Proceedings of the IEEE, 80(9), 1412–1434.
22. Clarkson, K. L., & Shor, P. W., (1989). Applications of random sampling in computational geometry, II. Discrete & Computational Geometry, 4(5), 387–421.
23. Crauser, A., Ferragina, P., Mehlhorn, K., Meyer, U., & Ramos, E. A., (2001). Randomized external-memory algorithms for line segment intersection and other geometric problems. International Journal of Computational Geometry & Applications, 11(3), 305–337.
24. Crauser, A., Ferragina, P., Mehlhorn, K., Meyer, U., & Ramos, E. A., (1999). I/O-optimal computation of segment intersections. External Memory Algorithms and Visualization, 4(2), 130–138.
25. Crowley, J., (1985). Navigation for an intelligent mobile robot. IEEE Journal on Robotics and Automation, 1(1), 31–41.
26. De Berg, M. T., Van, K. M., Overmars, M., & Schwarzkopf, O., (2000). Computational Geometry: Algorithms and Applications (Vol. 4, No. 2, pp. 5–9). Springer Science & Business Media.

27. De Berg, M., Cheong, O., Van, K. M., & Overmars, M., (2008). Line segment intersection: Thematic map overlay. Computational Geometry: Algorithms and Applications, 2(3), 19–43.
28. De Floriani, L., Magillo, P., & Puppo, E., (2000). Applications of computational geometry to geographic information systems. Handbook of Computational Geometry, 7, 333–388.
29. De Luna, M. L., Sato, A. K., Tsuzuki, M. D. S. G., Takimoto, R. Y., & De Castro, M. T., (2014). Intersection of Line Segments in a Discrete Domain, 2(5), 3–7.
30. Deans, S. R., (1981). Hough transform from the radon transform. IEEE Transactions on Pattern Analysis and Machine Intelligence, (2), 185–188.
31. Dévai, F. L., (1995). On the complexity of some geometric intersection problems. Journal of Computing and Information, 1(1), 333–352.
32. Devai, F., (1986). Quadratic bounds for hidden line elimination. In: Proceedings of the Second Annual Symposium on Computational Geometry (Vol. 2, pp. 269–275).
33. Devillers, O., & Fabri, A., (1993). Scalable algorithms for bichromatic line segment intersection problems on coarse-grained multicomputers. In: Workshop on Algorithms and Data Structures (Vol. 1, pp. 277–288). Springer, Berlin, Heidelberg.
34. Dobrindt, K., Mehlhorn, K., & Yvinec, M., (1993). A complete and efficient algorithm for the intersection of a general and a convex polyhedron. In: Workshop on Algorithms and Data Structures (Vol. 1, pp. 314–324). Springer, Berlin, Heidelberg.
35. Driscoll, T. A., (1996). Algorithm 756: A MATLAB toolbox for schwarz-Christoffel mapping. ACM Transactions on Mathematical Software (TOMS), 22(2), 168–186.
36. Dueker, K. J., (1987). Geographic information systems and computer-aided mapping. Journal of the American Planning Association, 53(3), 380–395.
37. Elliriki, M., Reddy, C. C. S., & Anand, K., (2019). An efficient line clipping algorithm in 2D space. Int. Arab J. Inf. Technol., 16(5), 798–807.
38. Fabri, A., & Pion, S., (2009). CGAL: The computational geometry algorithms library. In: Proceedings of the 17th ACM SIGSPATIAL International Conference on Advances in Geographic Information Systems (Vol. 1, 2, pp. 538, 539).
39. Fan, C., Luo, J., Liu, J., & Xu, Y., (2011). Half-plane Voronoi diagram. In: 2011 Eighth International Symposium on Voronoi Diagrams in Science and Engineering (Vol. 1, pp. 127–133). IEEE.

40. Fink, E., & Wood, D., (1996). Fundamentals of restricted-orientation convexity. Information Sciences, 92(1–4), 175–196.
41. Fogaras, D., & Sugihara, K., (2002). Topology-oriented construction of line arrangements. IEICE Transactions on Fundamentals of Electronics, Communications and Computer Sciences, 85(5), 930–937.
42. Fogel, E., Wein, R., & Halperin, D., (2004). Code flexibility and program efficiency by genericity: Improving CGAL's arrangements. In: European Symposium on Algorithms (Vol. 1, pp. 664–676). Springer, Berlin, Heidelberg.
43. Forrest, A. R., (1991). Computational geometry in practice. In: Fundamental Algorithms for Computer Graphics (Vol. 2, pp. 707–724). Springer, Berlin, Heidelberg.
44. Frankel, A., Nussbaum, D., & Sack, J. R., (2004). Floating-point filter for the line intersection algorithm. In: International Conference on Geographic Information Science (Vol. 1, pp. 94–105). Springer, Berlin, Heidelberg.
45. Gardan, Y., & Perrin, E., (1996). An algorithm reducing 3D Boolean operations to a 2D problem: Concepts and results. Computer-Aided Design, 28(4), 277–287.
46. Ghosh, S. K., & Mount, D. M., (1991). An output-sensitive algorithm for computing visibility graphs. SIAM Journal on Computing, 20(5), 888–910.
47. Goodrich, M. T., (1992). A polygonal approach to hidden-line and hidden-surface elimination. CVGIP: Graphical Models and Image Processing, 54(1), 1–12.
48. Guibas, L. J., & Seidel, R., (1987). Computing convolutions by reciprocal search. Discrete & Computational Geometry, 2(2), 175–193.
49. Güting, R. H., & Schneider, M., (1993). Realms: A foundation for spatial data types in database systems. In: International Symposium on Spatial Databases (Vol. 1, pp. 14–35). Springer, Berlin, Heidelberg.
50. Güting, R. H., & Schneider, M., (1995). Realm-based spatial data types: The ROSE algebra. The VLDB Journal, 4(2), 243–286.
51. Güting, R. H., (1994). An introduction to spatial database systems. The VLDB Journal, 3(4), 357–399.
52. Halperin, D., & Sharir, M., (2017). Arrangements. In: Handbook of Discrete and Computational Geometry (Vol. 2, pp. 723–762). Chapman and Hall/CRC.
53. Hanniel, I., & Halperin, D., (2000). Two-dimensional arrangements in CGAL and adaptive point location for parametric curves. In: International

Workshop on Algorithm Engineering (Vol. 3, pp. 171–182). Springer, Berlin, Heidelberg.

54. Held, M., (1997). Erit—A collection of efficient and reliable intersection tests. Journal of Graphics Tools, 2(4), 25–44.
55. Held, M., (2001). VRONI: An engineering approach to the reliable and efficient computation of Voronoi diagrams of points and line segments. Computational Geometry, 18(2), 95–123.
56. Hertel, S., & Mehlhorn, K., (1983). Fast triangulation of simple polygons. In: International Conference on Fundamentals of Computation Theory (Vol. 1, pp. 207–218). Springer, Berlin, Heidelberg.
57. Hillier, B., & Iida, S., (2005). Network and psychological effects in urban movement. In: International Conference on Spatial Information Theory (Vol. 1, pp. 475–490). Springer, Berlin, Heidelberg.
58. Hobby, J. D., (1999). Practical segment intersection with finite-precision output. Computational Geometry, 13(4), 199–214.
59. Hoel, E. G., & Samet, H., (1992). A qualitative comparison study of data structures for large line segment databases. In: Proceedings of the 1992 ACM SIGMOD International Conference on Management of Data (Vol. 1, pp. 205–214).
60. Hoel, E. G., & Samet, H., (1994). Data-parallel spatial join algorithms. In: 1994 International Conference on Parallel Processing (Vol. 3, pp. 227–234). IEEE.
61. Hoff, III. K. E., Keyser, J., Lin, M., Manocha, D., & Culver, T., (1999). Fast computation of generalized Voronoi diagrams using graphics hardware. In: Proceedings of the 26th Annual Conference on Computer Graphics and Interactive Techniques (Vol. 1, pp. 277–286).
62. Ilinkin, I., Janardan, R., Majhi, J., Schwerdt, J., Smid, M., & Sriram, R., (2001). A decomposition-based approach to layered manufacturing. In: Workshop on Algorithms and Data Structures (Vol. 1, pp. 389–400). Springer, Berlin, Heidelberg.
63. Imai, H., & Asano, T., (1984). Dynamic segment intersection search with applications. In: 25th Annual Symposium on Foundations of Computer Science (Vol. 1, pp. 393–402). IEEE.
64. Jokanovic, S., (2019). Two-dimensional line segment–triangle intersection test: Revision and enhancement. The Visual Computer, 35(10), 1347–1359.
65. Keeler, T., Fedorkiw, J., & Ghali, S., (2007). The spherical visibility map. Computer-Aided Design, 39(1), 17–26.

66. Kelly, D., (1987). Fundamentals of planar ordered sets. Discrete Mathematics, 63(2, 3), 197–216.
67. Kim, K., & Shan, J., (2011). Building roof modeling from airborne laser scanning data based on a level set approach. ISPRS Journal of Photogrammetry and Remote Sensing, 66(4), 484–497.
68. Kowaluk, M., & Majewska, G., (2015). $$\beta $$-skeletons for a set of line segments in $$ R^ 2$$ R2. In: International Symposium on Fundamentals of Computation Theory (Vol. 1, pp. 65–78). Springer, Cham.
69. Laidlaw, D. H., Trumbore, W. B., & Hughes, J. F., (1986). Constructive solid geometry for polyhedral objects. In: Proceedings of the 13th Annual Conference on Computer Graphics and Interactive Techniques (Vol. 1, pp. 161–170).
70. Lane, J. M., & Riesenfeld, R. F., (1980). A theoretical development for the computer generation and display of piecewise polynomial surfaces. IEEE Transactions on Pattern Analysis and Machine Intelligence, (1), 35–46.
71. Lasser, D., (1986). Intersection of parametric surfaces in the Bernstein-Bezier representation. Computer-Aided Design, 18(4), 186–192.
72. Lazarus, F., Pocchiola, M., Vegter, G., & Verroust, A., (2001). Computing a canonical polygonal schema of an orientable triangulated surface. In: Proceedings of the Seventeenth Annual Symposium on Computational Geometry (Vol. 3, pp. 80–89).
73. Li, Y., Wang, F., Stevenson, R., Fan, R., & Tan, H., (2018). Reliable line segment matching for multispectral images guided by intersection matches. IEEE Transactions on Circuits and Systems for Video Technology, 29(10), 2899–2912.
74. Li, Z., & Dong, W., (2010). A stroke-based method for automated generation of schematic network maps. International Journal of Geographical Information Science, 24(11), 1631–1647.
75. Li, Z., & Openshaw, S., (1992). Algorithms for automated line generalization1 based on a natural principle of objective generalization. International Journal of Geographical Information Systems, 6(5), 373–389.
76. Lu, C., Lin, Y., Ji, Z., & Dalian Univ. of Technology, (2005). An algorithm for plane-surface intersection and its application to shipbuilding. Ship Technology Research, 52(3), 103–106.
77. Magee, M. J., & Aggarwal, J. K., (1984). Determining vanishing points from perspective images. Computer Vision, Graphics, and Image Processing, 26(2), 256–267.

78. Mairson, H. G., & Stolfi, J., (1988). Reporting and counting intersections between two sets of line segments. In: Theoretical Foundations of Computer Graphics and CAD (Vol. 1, pp. 307–325). Springer, Berlin, Heidelberg.
79. Mehlhorn, K., Sharir, M., & Welzl, E., (1993). Tail estimates for the efficiency of randomized incremental algorithms for line segment intersection. Computational Geometry, 3(4), 235–246.
80. Miller, H. J., (2005). A measurement theory for time geography. Geographical Analysis, 37(1), 17–45.
81. Mittal, R., & Bhowmick, P., (2012). SPR e AD: On spherical part recognition by axial discretization in 4D Hough space. In: International Conference on Computer Vision and Graphics (Vol. 1, pp. 188–195). Springer, Berlin, Heidelberg.
82. Mortensen, C. W., (2003). Fully-dynamic two-dimensional orthogonal range and line segment intersection reporting in logarithmic time. In: Proceedings of the Fourteenth Annual ACM-SIAM Symposium on Discrete Algorithms (Vol. 1, pp. 618–627).
83. Newell, M. E., Newell, R. G., & Sancha, T. L., (1972). A solution to the hidden surface problem. In: Proceedings of the ACM Annual Conference (Vol. 1 pp. 443–450).
84. Ohdan, M., Ueno, Y., & Aizawa, K., (2014). An algorithm for counting line segments intersections using planar partitions. In: 2014 6th International Symposium on Communications, Control and Signal Processing (ISCCSP) (Vol. 2, pp. 133–136). IEEE.
85. Oyler, D. W., Kabamba, P. T., & Girard, A. R., (2014). Pursuit-evasion games in the presence of a line segment obstacle. In: 53rd IEEE Conference on Decision and Control (Vol. 2, pp. 1149–1154). IEEE.
86. Park, S. C., & Shin, H., (2002). Polygonal chain intersection. Computers & Graphics, 26(2), 341–350.
87. Pfoser, D., & Jensen, C. S., (1999). Capturing the uncertainty of moving-object representations. In: International Symposium on Spatial Database (4th edn., pp. 111–131). Springer, Berlin, Heidelberg.
88. Preparata, F. P., & Shamos, M. I., (2012). Computational Geometry: An Introduction (Vol. 1, pp. 2–5). Springer Science & Business Media.
89. Qi, D., Panneerselvam, K., Ahn, W., Arikatla, V., Enquobahrie, A., & De, S., (2017). Virtual interactive suturing for the fundamentals of laparoscopic surgery (FLS). Journal of Biomedical Informatics, 75, 48–62.

90. Quirk, J. J., (1994). An alternative to unstructured grids for computing gas dynamic flows around arbitrarily complex two-dimensional bodies. Computers & Fluids, 23(1), 125–142.
91. Razzazi, M., & Fotouhi, A., (2010). Coverage of wireless sensor networks in the presence of transparent obstacles. In: SoftCOM 2010, 18th International Conference on Software, Telecommunications and Computer Networks (Vol. 1, pp. 209–213). IEEE.
92. Requicha, A. A., & Voelcker, H. B., (1985). Boolean operations in solid modeling: Boundary evaluation and merging algorithms. Proceedings of the IEEE, 73(1), 30–44.
93. Rivero, M., & Feito, F. R., (2000). Boolean operations on general planar polygons. Computers & Graphics, 24(6), 881–896.
94. Rother, C., (2002). A new approach to vanishing point detection in architectural environments. Image and Vision Computing, 20(9, 10), 647–655.
95. Ruzon, M. A., & Tomasi, C., (2000). Alpha estimation in natural images. In: Proceedings IEEE Conference on Computer Vision and Pattern Recognition. CVPR 2000 (Cat. No. PR00662) (Vol. 1, pp. 18–25). IEEE.
96. Saalfeld, A., (1987). Technical note It doesn't make me nearly as CROSS Some advantages of the point-vector representation of line segments in automated cartography. International Journal of Geographical Information System, 1(4), 379–386.
97. Samet, H., & Webber, R. E., (1988). Hierarchical data structures and algorithms for computer graphics. I. Fundamentals. IEEE Computer Graphics and Applications, 8(3), 48–68.
98. Sarnak, N., & Tarjan, R. E., (1986). Planar point location using persistent search trees. Communications of the ACM, 29(7), 669–679.
99. Schoen, A. H., (1970). Infinite Periodic Minimal Surfaces Without Self-Intersections (No. C-98), 2(1), 3–7.
100. Shamos, M. I., & Hoey, D., (1976). Geometric intersection problems. In: 17th Annual Symposium on Foundations of Computer Science (sfcs 1976) (Vol. 2, pp. 208–215). IEEE.
101. Sherbrooke, E. C., Patrikalakis, N. M., & Brisson, E., (1996). An algorithm for the medial axis transform of 3D polyhedral solids. IEEE Transactions on Visualization and Computer Graphics, 2(1), 44–61.
102. Skala, V., (1989). Algorithms for 2D line clipping. In: New Advances in Computer Graphics (2nd edn., pp. 121–128). Springer, Tokyo.

103. Theobald, D. M., (2001). Topology revisited: Representing spatial relations. International Journal of Geographical Information Science, 15(8), 689–705.

104. Tilove, R. B., (1980). Set membership classification: A unified approach to geometric intersection problems. IEEE Transactions on Computers, 29(10), 874–883.

105. Vahrenhold, J., (2005). Line-segment intersection made in-place. In: Workshop on Algorithms and Data Structures (pp. 146–157). Springer, Berlin, Heidelberg.

106. Vaishnavi, V. K., & Wood, D., (1982). Rectilinear line segment intersection, layered segment trees, and dynamization. Journal of Algorithms, 3(2), 160–176.

107. Van, D. B. G., (2003). Collision Detection in Interactive 3D Environments (Vol. 3, No. 2, pp. 2–7). CRC Press.

108. Vassilev, T. S., & Eades, B., (2013). Generalizations of the Voronoi diagram. American Journal of Computational and Applied Mathematics, 3(2), 91–96.

109. Vogt, A., & Bared, J., (1998). Accident models for two-lane rural segments and intersections. Transportation Research Record, 1635(1), 18–29.

110. Wegman, E. J., (1990). Hyperdimensional data analysis using parallel coordinates. Journal of the American Statistical Association, 85(411), 664–675.

111. White, L. J., & Cohen, E. I., (1980). A domain strategy for computer program testing. IEEE Transactions on Software Engineering, 1(3), 247–257.

112. Wong, K., & Müller, H. A., (1991). An Efficient Implementation of Fortune's Plane-Sweep Algorithm for Voronoi Diagrams (Vol. 1, No. 4, pp. 5–9). University of Victoria, Department of Computer Science.

113. Wu, H., Gong, J., Li, D., & Shi, W., (2000). An algebraic algorithm for point inclusion query. Computers & Graphics, 24(4), 517–522.

114. Xu, G., & Zhang, Z., (1996). Epipolar Geometry in Stereo, Motion and Object Recognition: A Unified Approach (Vol. 6, pp. 1–5). Springer Science & Business Media.

115. Yang, M. C., & Yang, C. C., (1984). Assessment of the intersection probabilities of random line segments and squares. Computer Vision, Graphics, and Image Processing, 26(3), 319–330.

116. Yi, J., (2012). An efficient occlusion culling algorithm of line segment intersection based on large-scale scene. In: 2012 8th International Conference on Information Science and Digital Content Technology

(ICIDT2012) (Vol. 1, pp. 128–130). IEEE.

117. Zhang, L., & Ghosh, B. K., (2000). Line segment based map building and localization using 2D laser rangefinder. In: Proceedings 2000 ICRA. Millennium Conference; IEEE International Conference on Robotics and Automation; Symposia Proceedings (Cat. No. 00CH37065) (Vol. 3, pp. 2538–2543). IEEE.
118. Zhou, Q., Zhong, E., & Huang, Y., (2009). A parallel line segment intersection strategy based on uniform grids. Geo-Spatial Information Science, 12(4), 257–264.

CHAPTER

4 Computational Geometry and Linear Programming

Contents

4.1. INTRODUCTION

Computational geometry has produced a large number of advanced algorithms for geometric challenges in low dimensions, all of which are based on the unified perspective of geometric algorithms that are used to approach the issues (Imai, 1991). A common misconception is that high-dimensional geometric problems are simply special cases of mathematical programming problems. However, this is not always the case. It goes without saying that when applied to problems with low dimensions, computational-geometric algorithms outperform broad algorithms for mathematical programming in terms of speed and efficiency (Figure 4.1) (Dyer et al., 2017).

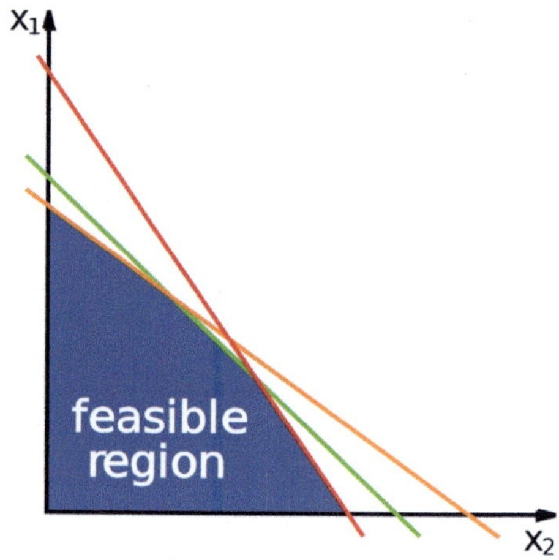

Figure 4.1. Identification of feasible region using linear programming.

Source: https://www.toppr.com/guides/maths/linear-programming/linear-programming-problem/.

For instance, a linear programming issue in a stable dimension can be fixed in linear time, whereas only minimally polynomial-time algorithms are known for solving a general linear programming problem (Yao, 1990). Though methods in computational geometry and mathematical programming have been combined in a limited number of studies, both fields should be investigated further in the future from the perspectives of their respective fields (Goldfarb and Todd, 1989). Several different optimizations for linear programming in computational geometry, like the randomization and prune-and-search paradigm, as well as results in mathematical programmings, such as the interior-point method, are discussed in this chapter, which begins with a survey of the results for linear programming in computational geometry (Seidel, 1991). A further demonstration is given on how to combine computational geometry and mathematical programming techniques to solve a variety of geometric difficulties in both high and low dimensions. Geometric algorithms, both parallel and sequential, that are both efficient and fast, are discussed.

Linear programming is an optimization challenge in which the objective function is minimized while being constrained by linear inequality constraints:

$$\min c^T x$$

$$\text{s.t. } Ax \geq b$$

where; c, x R^d, b R^N, A R^{Nxd} A, b, c are given, and elements x_1, x_d of d are x variables. Linear programming is the ideal optimization model for real-world applications because it can be used to simulate a wide range of discrete and continuous systems, and it can solve large-scale linear programming problems with thousands of variables or more in a reasonable amount of time (Borrelli et al., 2003).

The wide application of linear programming does research into linear programming methods in the branch of computer science particularly intriguing. Linear programming contains a lot of interesting characteristics in terms of algorithm theory, such as the following.

For starters, linear programming may be thought of as both discrete and continuous optimization. The issue is a continuous approach to solve in its original form. By looking at the feasible region x Axb, which is a curved polyhedron, it can be seen that if rank A = d, there occurs an ideal solution that is an extreme point of the curved polyhedron, and since the number of extreme points of A is constrained by. , the problem can be solved in principle by verifying all the extreme points in a distinct manner. Dobkin devised a linear programming technique called the simplex method that travels a path of contiguous extreme points of the polyhedron to an optimal position based on this distinct characteristic. In the business realm, machine learning is an application that has been shown to be a beneficial tool (Dobkin, 1992).

Second, linear programming's time complexity is quite close to the border between tractable and intractable problems. Linear programming is registered as one of some big challenges that were not famous for falling into which of the intractable or tractable classes at the time in a renowned book on NP-comprehensiveness by researcher. In fact, in the worst-case scenario, the simplex approach may navigate all of the convex polyhedron's extreme points. The simplex system is considered a non-polynomial system overall because the number of extreme points can be exponential. Edelsbrunner demonstrated in 1979 that linear programming could be resolved in polynomial time using an ellipsoid technique (Edelsbrunner et al., 1984). Because the inside of the polyhedron has a continuous structure, the interior-point technique can be described as a linear programming algorithm with a continuous structure (Berg et al., 1997).

Third, no strongly polynomial-time algorithms are known to exist in linear programming. Strongly polynomial-time algorithms have already been developed for system flow challenges, which are a valuable subclass of linear programming, and this problem has been a big open issue regarding the difficulty of linear programming (Lee and Preparata, 1984).

Fourth, in a few applications, like computer visuals, linear programming problems emerge in which 'd' is significantly $<$ n or 'd' is a constant, such as three, which are difficult to solve. It is possible to deal with this sort of linear programming issue in computational geometry, which is a branch of mathematics. Computational geometry, which was coined by Shamos in the mid-1970s is a branch of computer science that approaches geometric issues in a cohesive manner from the standpoint of algorithms.

As a consequence of the field's efforts, efficient methods for constructing the convex hull, Voronoi diagram, arrangement, and other geometric figures have been devised. For a long time, computational geometry has focused on low-dimensional geometric issues, particularly those that occur in two- or three-dimensional space (Greene and Yao, 1986).

Furthermore, the vast majority of computational geometric methods are discrete in nature. From this point forward, computational geometry should be applied to higher-dimensional challenges and should employ more continuous techniques. Randomization is one of the beneficial approaches that computational geometry is presently employing, and it has some link to continuous structure in some way.

It is because of this those algorithms begin to exhibit probabilistic behavior, and it is necessary to consider geometric issues in a more generic environment in order to study such randomized algorithms. When the dimension is treated as a constant in linear programming, computational geometry produces a linear-time method that can be implemented. The prune-and-search concept is employed in order to derive such an algorithm. Originally, this paradigm was employed in order to produce a linear-time algorithm for algorithm selection. The prune-and-search theory is extended to higher-dimensional difficulties through the use of computational geometry, and this, in addition to linear programming, results in the development of several useful algorithms (Figure 4.2) (Overmars, 1988).

Figure 4.2. Assumptions in linear programming.

Source: https://businessjargons.com/assumptions-of-linear-programming.html.

In this chapter, we will go through some of the outcomes that have been discussed before in relation to linear programming. Due to the fact that this is being done from the author's perspective, the issues covered may be prejudiced when contrasted to the paper's title, which is far too broad in scope. By emphasizing such achievements, which would be of practical significance in and of themselves, we want to provide some insight into possible future research directions in the associated domains in the future. Despite the fact that linear programming has been emphasized up to this point in the introduction from the perspective of computational geometry, mathematical programming will be treated as a major subject in this chapter (Sethian, 1999).

The prune-and-search model, as well as its applicability to the two-dimensional linear programming question, are discussed in length in the following section. Following that, two examples of the prune-and-search approach being used to solve certain particular linear systems are discussed. We also discuss randomized techniques for linear programming, which are described in detail below.

Automated manufacturing is used to create the vast majority of the items we see around us now, from automobile bodies to plastic cups and crockery to household items. This is a process in which computers play a significant role, both during the design phase and during the building phase; CAM/CAD capabilities are an essential element of any contemporary manufacturing. Factors such as the material to be used in manufacturing the product, its shape, and whether or not the object will be mass-produced all influence which manufacturing technique is utilized to create a certain object (Weishan et al., 1997). In this chapter, we will look at some of the geometric features of manufacturing utilizing molds, which is a procedure that is widely used to create plastic or metal things. Casting is a term used to describe the process of creating metal items (Thomas, 2018).

The casting process is seen in Figure 4.3; liquid metal is rained into a mold, it hardens, and the item is subsequently withdrawn from the mold. The last stage is not always as simple as it appears; the object may become caught in the mold and be impossible to remove without ruining it. We can sometimes get around this issue by utilizing a different mold. However, there are some objects for which no suitable mold exists, such as a spherical (Rajagopal, 1981). This is the issue we'll look at in this chapter: is there a mold for an object that can be used to remove it?

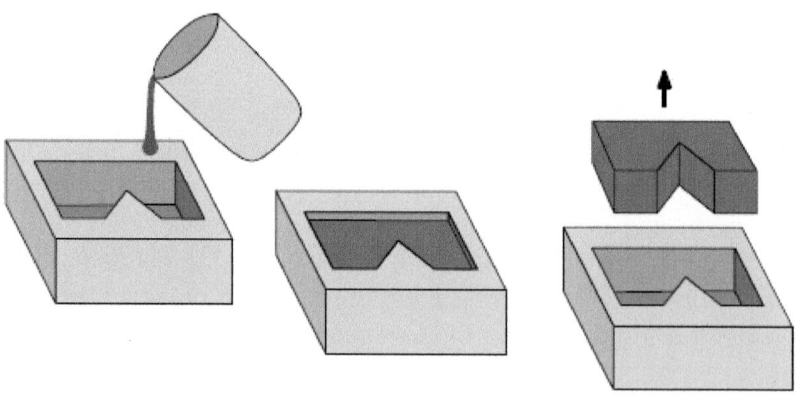

Figure 4.3. The casting processes.

Source: https://www.amazon.com/Computational-Geometry-Applications-Mark-Berg/dp/3540779736.

We'll stick to the following scenario for now. For starters, we'll suppose the item to be built polyhedral. Second, we only examine one-piece molds, not two-piece or multi-piece molds (Smaoui et al., 2004). (By employing two-piece molds, it is feasible to make things such as spheres that would be impossible to make with a single-piece mold.) Lastly, we only allow a single translation to extract the item from the mold. We won't be able to extract a bolt from its mold as a result of this. For many items, translational movements are sufficient (Watanabe et al., 1997).

4.2. THE GEOMETRY OF CASTING

If we want to see if an object can be cast, we need to identify an appropriate mold for it. The form of the item determines the shape of the cavity in the mold; however, varied angles of the object result in different molds. Selecting the right orientation is important since some orientations result in molds that are impossible to remove, while others allow the item to be removed. The object must have a top horizontal facet, which is an evident limitation on the orientation (Wonisch et al., 2011). This is the only aspect that will not come into touch with the mold. As a result, there are as several alternative orientations—or, equally, molds—as there are facets on the item. If an item can be released from its mold for at least one of these orientations, we call it castable. The next sections will focus on establishing if an object may be removed using a translation from a certain mold. We just attempt every possible orientation to determine the object's castability (Figure 4.4) (Lawler et al., 2011).

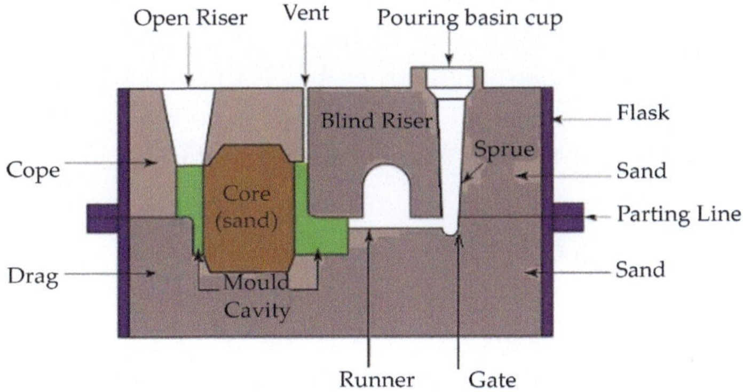

Figure 4.4. Casting mold geometry.

Source: https://www.theengineerspost.com/metal-casting-process-types/.

Let 'P' be a 3D polyhedron with a designated top facet—that is, a 3D solid bounded by planar aspects. We'll pretend the mold is a rectangular building block with a hollow that perfectly matches P. When the polyhedron is inserted in the mold, its top facet should be parallel to the xy-plane and coplanar with the topmost facet of the mold. This indicates that there are no extraneous pieces protruding from the top of the mold that may prevent 'P' from being separated (Chen et al., 1987).

A component of 'P' that is not the top facet is referred to as a common aspect. All ordinary facet 'f' in the mold has a matching facet, which we call f.

We're trying to figure out if 'P' can be freed from its mold with just one translation. To put it another way, we want to know if there is a direction 'd' in which 'P' may be translated to eternity in direction 'd' devoid of crossing the interior of the mold. It's worth noting that we let 'P' move down the mold. Because the top facet of 'P' does not touch the mold, the removal direction must be upward, resulting in a positive z-component. This is just a required situation on the elimination direction; other restrictions are required to ensure that a direction is legitimate (Onishi and Takayama, 1996).

Assume that 'f' is a regular aspect of P. This facet must slip away from or along with the mold's matching facet f. We must define the angle of two vectors in 3-space to make this restriction accurate. This is how we go about it.

The angle of the vectors is the lesser of the two angles measured in the plane spanned by the vectors. With. , f's outward normal, 'f' now prevents any translation in a direction that makes an angle of $< 90°$. As a result, one of the required properties of 'd' is that it forms a 90° angle with the outward normal of every normal aspect of P. This condition is also adequate, as shown by the next theorem (Wehr and Woo, 1998).

4.3. HALF-PLANE INTERSECTION

Let $H = h_1, h_2, \ldots, h_n$ denote a collection of linear constraints in two variables, i.e., constraints of the type aix+biy ci, where ai, bi, and ci are constants with at least one of ai and bi being non-zero. A closed half-plane in R_2 bordered by the line aix+biy = ci can be interpreted geometrically as such a restriction. In this part, we'll look at how to identify the set of all (x,y) R_2 points that meet all 'n' criteria at the same time. To put it another way, we're looking for all the points in 'H' that are at the common intersection of the half-planes

(Masajada, 2000). A half-plane is curved, and the interconnection of curved sets is another convex set, so the interconnection of a set of half-planes is a convex region in the plane. Every point on the crossing boundary must be on a half-bounding plane's line. As a result, the region's boundary is made up of edges contained within these bounding lines. Due to the convexity of the intersection, each bounding line can only contribute one edge (Fan et al., 2002). The junction of 'n' half-planes is, therefore, a convex polygonal region with at most 'n' edges. Figure 4.5 depicts a few instances of half-plane intersections. Dark shading in the figure indicates which side of its bounding line a half-plane lies on; the common intersection is shaded lightly (Lupu, 2016). The interconnection does not have to be bounded, as shown in Figures 4.5(ii) and (iii). As in (iv), the connection can disintegrate into a line segment, or it can be empty (v) or a point.

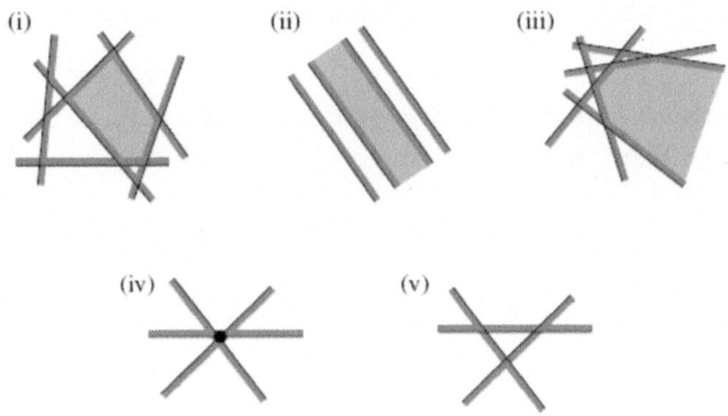

Figure 4.5. Illustrations of the junction of half-planes.

Source: https://slidetodoc.com/half-plane-intersections-let-h-h-1-h/.

To calculate the interconnection of a set of 'n' half-planes, we present a simple divide-and-conquer algorithm. It uses the INTERSECTCONVEXREGIONS routine to find the connection between the two curved polygonal regions. First, we'll go over the whole algorithm.

- **Algorithm INTERSECTHALFPLANES(H):**

Input: A set 'H' of 'n' half-planes in the plane.

Output: The convex polygonal region C:= $\bigcap_{h \in H} h$.

- if card(H) = 1;
- then C ← the exclusive half-plane h $\in$ H;
- else Separate 'H' into sets H_1 and H_2 of size $\lceil n/2 \rceil$ and $\lfloor n/2 \rfloor$;

- C ←INTERSECTCONVEXREGIONS(C_1, C_2);
- C_1 ←INTERSECTHALFPLANES(H_1);
- C_2 ←INTERSECTHALFPLANES(H_2).

The process INTERSECTCONVEXREGIONS still needs to be described. But wait, wasn't this issue addressed in Chapter 2? Indeed, according to Corollary 2.7, the intersection of two polygons may be computed in O(nlogn + klogn) time, where 'n' is the total number of vertices in both polygons. When applying this solution to our issue, we must be cautious since the areas we have might be unbounded into a segment or a point (Lawler et al., 2002).

4.4. PRUNE-AND-SEARCH MODEL AND ITS APPLICATION TO 2D LINEAR PROGRAMMING

Several algorithmic models handle smaller subproblems of a larger problem to arrive at a solution to the overall challenge. This is due to the fact that the smaller the problem, the easier it is to solve (McCormack and McNamara, 2000). The prune-and-search model, which is applied to the 2D system of equations below, aims to reduce the size of the feature by a constant factor by deleting unnecessary items at every stage (Figure 4.6) (Masajada, 2000).

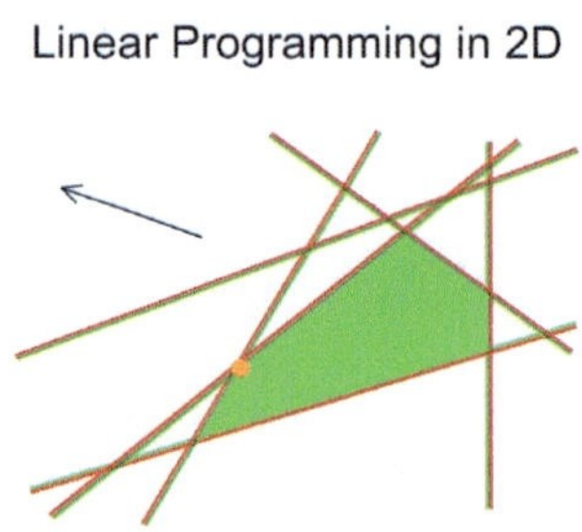

Figure 4.6. Illustration of 2D linear programming.

Source: https://www.slideserve.com/zaynah/linear-programming-in-low-dimensions.

In the design and study of algorithms, the prune-and-search concept is one of the most helpful paradigms. It's been utilized in linear-time selection methods before (Aghajani and Moradifam, 2006). As previously stated, one of the core concepts of this model is to delete unnecessary pieces by a fixed factor at every repetition. The algorithm discovers a subset of si elements that are either all < x or all > x at the ith iteration when choosing the k(= ko)th element 'x' including 'n' elements. Then, for ki = ki1 – si and ki1 based on whether these elements in the small segment are < or < than x, find kith element among the organic part in the next step (Lawler et al., 2011). Obtaining a subset of items whose size is assumed to be no less than a constant factor a1 of the present size can be done in time linear to the current size, roughly speaking. After that, the total time complexity is bounded in magnitude by:

$$n+(1-\alpha)n+(1-\alpha)^2 n+\cdots\leq\frac{1}{\alpha}n$$

A linear-time algorithm is therefore achieved (O'Rourke et al., 1982).

Let's look at how Megiddo et al. used the prune-and-search model to construct a linear-time method for the 2D linear programming issue. We aim to offer a fairly detailed summary of this algorithm because it is relatively straightforward to discuss compared to the other approaches in this study (McNamara et al., 2000). The following is a typical 2D linear programming issue with 'n' inequality restrictions:

$$\min\ c_1x_1+c_2x_2$$

$$\text{s.t. } a_{i1}x_1+a_{i2}x_2\geq a_{i0} \qquad (i=1,\ldots\ldots\ldots\ldots n)$$

Identifying the feasible region fulfilling the inequality requirements on the (x_1, x_2)-plane, which is just a curved polygon if limited, would make solving the problem illustratively quite straightforward. Rather than considering the topic as its whole, we will focus our attention on the following specific difficulty (Figure 4.7) (Bueckner, 1987).

$$\min\ y$$

$$\text{s.t. } y\geq a_ix+b_i \qquad (i=1,\ldots\ldots\ldots\ldots n)$$

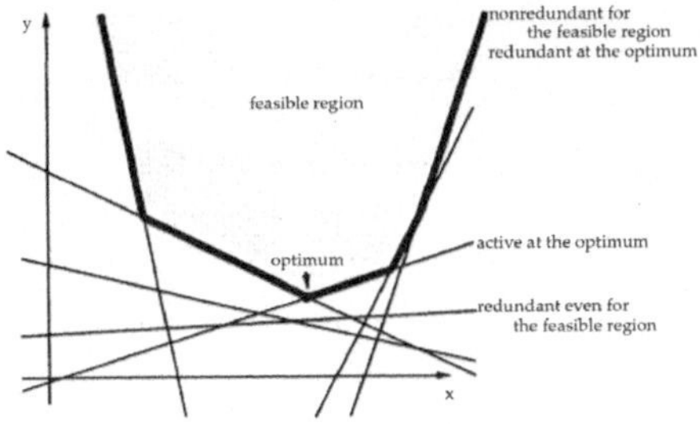

Figure 4.7. A two-dimensional linear programming problem.

Source: https://www.researchgate.net/figure/Graphic-representation-of-the-elements-of-a-generic-two-dimensional-linear-optimization_fig3_334852870.

This is because this particular problem is nearly adequate for devising a linear-time method for the broader 2D problem, and its simpler structure is better for demonstrating the prune-and-search technique's essence (Chen and Xu, 2002). This n = 7 issue is depicted in Figure 4.7

We regard the problem as described above. Define a function f(x) by:

$$f(x) = \max\{a_i x + b_i \mid i = 1, \ldots\ldots\ldots, n\}$$

The task is therefore reduced to minimizing f. (x). In Figure 4.8, the graph of y = f(x) is displayed in bold lines. f(x) has the following good attribute (De et al., 2015). First, we'll go through what it means for a function to be convex. If the function g: 'R R' is convex, then.

$$g(\lambda x_1 + (1-\lambda)x_2) \leq \lambda g(x_1) + (1-\lambda)g(x_2)$$

For any x_i, x_2, A ∈ R with 0 < A < 1. If the above inequality constantly stands strictly, g(x) is called strictly convex (Bhattacharya et al., 2020).

Nonetheless, this application of the prune-and-search methodology to 2D linear programming is quite like linear-time selection (Agarwal and Sharir, 1998). When the prune-and-search approach is extended to higher dimensions, the algorithms that result are really computational-geometric. The following is how the higher-dimensional prune-and-search strategy works.

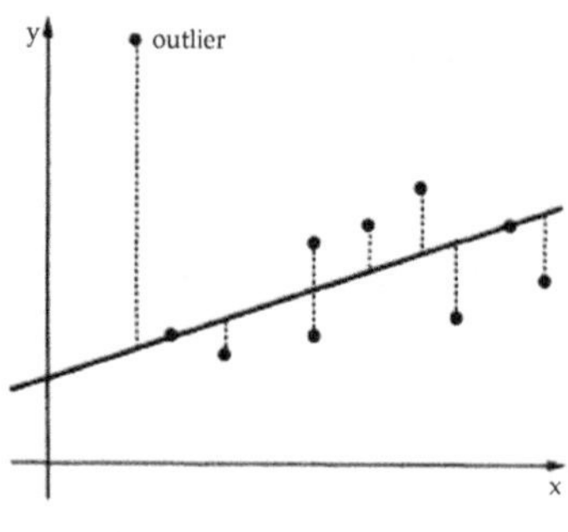

Figure 4.8. Linear L1 estimate of points.

Source: https://www.kurims.kyoto-u.ac.jp/~kyodo/kokyuroku/contents/pdf/0872-06.pdf.

The general approach assumes that an ideal solution is defined as a maximum of a fixed number of objects, which in the nondegenerate case of linear programming is 'd' (Panigrahi et al., 2013). Most of the time, the algorithm searches in the d-dimensional space relative to 'n' objects by iteratively solving a constant number of sub-problems in the (d-1)-dimensional space, thereby reducing the size of the problem in the d-dimensional space by a constant factor (Figure 4.8) (Daescu and Palmer, 2009). The method prunes the remaining items by a constant factor, a, at each iteration by performing a test a constant number of times (Buzer, 2003). Because the complexity of the method is dependent on the test being willing to describe the relative location of the best solution in linear time with regard to the number of residual objects, the test in d-dimensional space is an essential component of the algorithm. The test in d-dimensional space is carried out by solving the (d — 1)-dimensional subproblems previously indicated. As a result, there are O(log n) steps in total, with the amount of time spent at every step decreasing geometrically as mentioned above, consuming linear time in total (Späth and Watson, 1987). Megiddo et al. (1997) were the first to use this method. Linear programming in a fixed dimension may be solved in $0(2^{2d})$ time, which is linear in 'n,' using this method (Abdelmalek, 1974). However, because the algorithm's time complexity is twice exponential in d, it may only be viable for small d. This difficulty has been reduced in a number of ways, but it is still exponential in terms of d. We'll come back to this topic in the following part. Before that, we'll go through two examples of how the prune-and-search approach may be employed to solve bigger special linear programming problems (Yamamoto et al., 1988).

4.5. LINEAR L1 APPROXIMATION OF POINTS

The line-fitting problem, which involves approximating a collection of 'n' points with a linear function or a line in the plane, is crucial in numerical computation and statistics (Houle et al., 1993). The least-squares approach is the most commonly employed; however, there are other approximation methods such as the L1 and Lc." approximation methods. The L1 approximation, in particular, is more resistant to outliers than the least-squares technique and is hence preferred for noisy data (Watson, 1981). Let 'S' be a set of 'n' points pi in the plane, and (x_i, y_i) I = 1, 'n' be the (x, y)-coordinate of point pi. The L1 approximation of an estimated line defined by y = axe + b with constraints 'a' and 'b' is defined by the subsequent error criteria (Robers and Ben-Israel, 1969), which minimizes the L1 norm of the estimate line to the point set S:

$$\min_{a,b} \sum_{i=1}^{n} | y_i - (ax_i + b) |$$

Figure 4.7 illustrates the case of n = 10 with an outlier.

This challenge can be created as the subsequent linear programming problem with n + 2 variables a, b, ci (i = 1, …, n):

$$\min \sum_{i=1}^{n} c_i$$

$$\text{s.t. } y_i - (ax_i + b) \leq c_i$$

$$-y_i + (ax_i + b) \leq c_i$$

The constants x_i, y_i I = 1, 'n' are provided below (Barrodale, 1968). Because there are n + 2 variables and additional inequalities in this linear programming issue, the linear-time solution for linear programming in a specified size cannot be used (Gurwitz, 1990). The issue, though, is fundamentally a 2D one. We may change this issue into a two-dimensional prune-and-search predicament by applying the point-line duality transformation, which is one of the most used techniques in computational geometry (Barrodale and Roberts, 1973). However, in the L1 approximation issue, each infinitesimal movement of any point in 'S' affects the norm (and perhaps the solution); hence there are no redundant points in terms of an optimum solution. As a result, applying the pruning strategy directly does not result in a linear-time algorithm (Cadzow, 2002).

Researchers provide a way for solving this problem by utilizing the spline linearity of the L1 norm to produce a linear-time algorithm (Yadav et al., 2021). Kato also applies this approach to higher-dimensional L1 approximation problems in his master's thesis. Despite their complexity, these algorithms demonstrate the strength of the prune-and-search approach in purely computational-geometric scenarios (Avron et al., 2010).

4.6. A RANDOMIZED ALGORITHM FOR LINEAR PROGRAMMING

This approach, known as prune-and-search, generates algorithms for linear programming in a fixed measurement that run in linear time, which is considered to be the best feasible result theoretically (Ascher, 1978). Although the time complexity increases exponentially with increasing dimension d, this does not prevent the methods from being inefficient in practice, even for 'd' of moderate magnitude. One method of overcoming this challenge is to employ randomization, which has long been acknowledged as a valuable technique in the field of computational geometry (Koh et al., 2007). This does not imply that any probability dispersal on the problem, such as the inequality restrictions in this example, is assumed to be true in this context. Randomization, on the other hand, incorporates probabilistic behavior into the process of algorithm development (Ray et al., 1993). When issues are randomly generated, it becomes likely to analyze the average case difficulty of problems in addition to the worst-case complexity, which is highly useful from a user standpoint. Furthermore, it is frequently the case that randomized algorithms are straightforward and straightforward to apply.

In this section, we will first provide a brief explanation of a randomized approach for linear programming developed by Clarkson. Take into consideration the problem of 2D linear programming that was discussed in Section 2.1. As a refresher, the problem may be thought of as locating two active inequality limitations at an optimal solution and then deleting all the other unnecessary constraints from the solution (Abdelmalek, 1985).

Take a subset So of constraints from a larger set of 'n' constraints at random and in isolation. Calculate an optimum solution (xo, yo) for this subproblem by solving a linear programming problem with this subset So of restrictions employing the similar objective function as in the previous subproblem (van den Brand et al., 2021). If all of the other rt — requirements are met at this optimal solution, then this optimal solution for the subset of constraints is also an optimal solution for all of the limits, and we are through with this problem (Andoni and Onak, 2012).

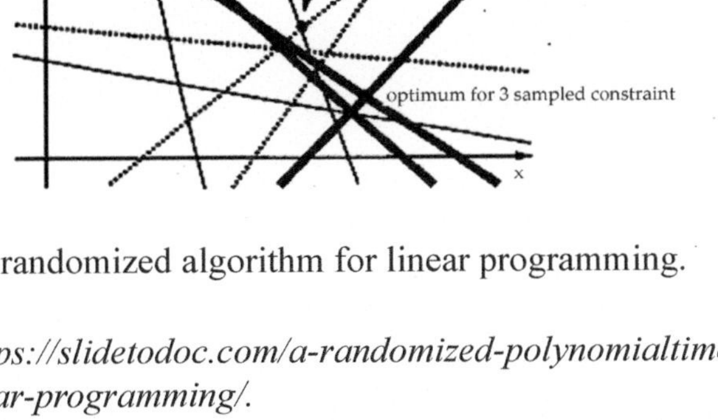

Figure 4.9. A randomized algorithm for linear programming.

Source: https://slidetodoc.com/a-randomized-polynomialtime-simplex-algorithm-for-linear-programming/.

We then calculate a set S_1 of constraints 'I' that is in violation of the computed solution: yo aixo + bi, unless otherwise specified. Figure 4.9

depicts the scenario of n = 9 in which three limitations are chosen arbitrarily, and there are constraints that violate the optimum for the sampled limitations (Appa and Smith, 1973) (as indicated by the dotted lines).

In Figure 4.7, Si contains precisely one constraint that is active at the worldwide optimum. The size of Si is determined by the random selection of a subset of constraints. The expected size of Si can be shown to be 0(in-), which is a crucial feature of this randomized algorithm. That is, we can get a set Si of limitations of likely size O(fn) that includes no less than one of active restrictions at the worldwide optimum solution by arbitrarily sampling a subset So of VT/ constraints (Dyer and Frieze, 1989).

We sample another set of Ft constraints, this time in S_i U S_2, and solve a linear programming problem with Si U S_2 constraints. Let S_3 be a set of limitations that violate the optimum solution to the S_i U S_2 subproblem. It can be seen that the likely size of S_3 is O(,/) and that at the global optimum solution, S_i U S_3 contains at least two active constraints. The initial issue for 'n' constraints is now decreased to that for O(n) limitations because Si U 53 includes those two active limitations at the ideal solution and its average size is 0(jii). The problem is then efficiently solved by recursively applying this procedure (Chan, 2004).

This is accomplished through the algorithm's use of random sampling. However, there is still an exponential in 'd' term in the complexity function, and the algorithm is not strongly polynomial (Ludwig, 1995).

The estimated number of breaching constraints to the optimal for the sampled set So is an important consideration in the analysis and design of this randomized algorithm (Karp, 1991). This evaluation can be completed entirely in a discrete manner in this case. However, in a more general instance, as in Bartal et al. (2000), the continuous model of probability may be applied, and adding make random in geometric algorithms may go-ahead to a greater investigation of continuous structures of geometric challenges (Matoušek et al., 1996).

Return to the previous subsection's assignment issue, which had k n supply vertices demand and vertices. As previously indicated, the prune-and-search model provides the 0((k+1)!)2n)-time algorithm. Seidel offers a randomized method with randomized time complexity O(kn + k3 – 5n°•5 log n) that uses random sampling. This approach is best for k n because the complexity is reduced to O(kn): As a result, randomization has a dramatic effect on the assignment issue with k n. This is made feasible by providing a decent bridge between combinatorial and geometry optimization, as well as by using the randomization model that is appropriate for geometric challenges (Seidel, 1990).

By merging computational geometric results with linear programming and its particular case, these randomized algorithms are regarded as beneficial results. After discussing a continuous strategy, we'll return to this topic in section 4.

4.7. LINEAR PROGRAMMING IN MATHEMATICAL PROGRAMMING

For large-scale linear programming issues in mathematical programming, a nonlinear technique has lately been proved to be effective, and the supposed interior-point technique is now regarded as a strong technique for linear programming (Bansal et al., 2012). We discuss the multiplicative penalty function approach suggested by Iri and Irnai in this section, which is one of many interior-point methods. Then, using the generic interior-point technique, we explain how to enhance an algorithm for the planar minimum-cost flow issue, which is a "unique example of linear programming." Another advantage of the interior-point approach for linear programming is this. The use of the interior-point approach to the construction of parallel linear programming algorithms is also examined.

4.8. MULTIPLICATIVE PENALTY FUNCTION TECHNIQUE FOR LINEAR PROGRAMMING

Researchers presented the multiplicative penalty function technique, which is an internal approach that minimizes the convex multiplicative penalty function provided by the Newton method for a given linear programming challenge with inequality limitations (Agarwal and Sen, 2001). In this view, the multiplicative penalty function approach is the easiest and most natural way among penalty function interior-point techniques. The multiplicative sentence function may be thought of as an affine form of Karmarkar's potential function. The method's local quadratic convergence was demonstrated, but the global convergence feature was left open. Under the premise that the line search can be conducted rigorously, Agarwal verify the method's global linear convergence. Imai shows that the multiplicative penalty function technique has O(n2L) main iterations, with each iteration's main purpose being to resolve a linear system of 'd' variables to obtain the Newton path. Researcher demonstrates that the unique multiplicative consequence function technique takes 0(n1.5L) main iterations, whereas an approach with a higher penalty parameter takes O(nL) main reiterations. In that study, it is also demonstrated that the multiplicative penalty function approach can solve the convex quadratic programming problem with the same complexity (Adler and Shamir, 1993).

For linear programming, there are currently a plethora of interior-point methods. In the manner described in the introduction, we investigate the following linear programming issue. We also assume the following in the sequel:

- The central Int X of the possible region X is filled;
- The viable region X = {x I Ax > b} is bounded;
- The minimum value of cTx is zero.

As a result, the objective is simplified to minimize f. (x). The graph of y = f(x) is shown in bold lines in Figure 4.7 The following is a positive feature of f(x). We'll start by discussing what it means for a function to be convex. If g: R R is a convex function, then:

$$F(x) = {(c^T x)^{n+\delta}} \Big/ {\prod_{i=1}^{n} (a_i^T x - b_i)}$$

where; ai^T ∈ R^d is the i-th row vector of A and 6 is a constant > or equal to 1. This function is first mentioned in a study, where the value of 6 is set to one. 6 is set to about n. The outlines of F(x) in the 2D situation with n = 4 are illustrated in Figure 4.7. When F(x) = 0, the distance between 'x' and the set of optimal solutions touches to zero under these assumptions. Starting from an initial interior point, the multiplicative penalty function technique uses the Newton method to directly minimize the penalty function F(x) (Pariyani et al., 2006).

REFERENCES

1. Abdelmalek, N. N., (1974). On the discrete linear L1 approximation and L1 solutions of overdetermined linear equations. Journal of Approximation Theory, 11(1), 38–53.
2. Abdelmalek, N. N., (1985). Piecewise linear L1 approximation of planar curves. International Journal of Systems Science, 16(4), 447–455.
3. Adler, I., & Shamir, R., (1993). A randomized scheme for speeding up algorithms for linear and convex programming problems with high constraints-to-variables ratio. Mathematical Programming, 61(1), 39–52.
4. Agarwal, P. K., & Sen, S., (2001). Randomized algorithms for geometric optimization problems. Combinatorial Optimization-Dordrecht, 9(1), 151–187.
5. Agarwal, P. K., & Sharir, M., (1998). Efficient algorithms for geometric optimization. ACM Computing Surveys (CSUR), 30(4), 412–458.
6. Aghajani, A., & Moradifam, A., (2006). Some sufficient conditions for the intersection with the vertical isocline in the Liénard plane. Applied Mathematics Letters, 19(5), 491–497.
7. Andoni, A., & Onak, K., (2012). Approximating edit distance in near-linear time. SIAM Journal on Computing, 41(6), 1635–1648.
8. Appa, G., & Smith, C., (1973). On L1 and Chebyshev estimation. Mathematical Programming, 5(1), 73–87.
9. Ascher, U., (1978). On the invariance of the interpolation points of the discrete L1-approximation. Journal of Approximation Theory, 24(1), 83–91.
10. Avron, H., Sharf, A., Greif, C., & Cohen-Or, D., (2010). ℓ1-sparse reconstruction of sharp point set surfaces. ACM Transactions on Graphics (TOG), 29(5), 1–12.
11. Bansal, N., Buchbinder, N., & Naor, J., (2012). A primal-dual randomized algorithm for weighted paging. Journal of the ACM (JACM), 59(4), 1–24.
12. Barrodale, I., & Roberts, F. D., (1973). An improved algorithm for discrete l_1 linear approximation. SIAM Journal on Numerical Analysis, 10(5), 839–848.
13. Barrodale, I., (1968). L1 approximation and the analysis of data. Journal of the Royal Statistical Society: Series C (Applied Statistics), 17(1), 51–57.
14. Bartal, Y., Chrobak, M., & Larmore, L. L., (2000). A randomized algorithm for two servers on the line. Information and Computation, 158(1), 53–69.
15. Berg, M. D., Kreveld, M. V., Overmars, M., & Schwarzkopf, O., (1997). Computational geometry. In: Computational Geometry (pp. 1–17). Springer, Berlin, Heidelberg.

16. Bhattacharya, B. K., De, M., Nandy, S. C., & Roy, S., (2020). Constant work-space algorithms for facility location problems. Discrete Applied Mathematics, 283, 456–472.
17. Borrelli, F., Bemporad, A., & Morari, M., (2003). Geometric algorithm for multiparametric linear programming. Journal of Optimization Theory and Applications, 118(3), 515–540.
18. Bueckner, H. F., (1987). Weight functions and fundamental fields for the penny-shaped and the half-plane crack in three-space. International Journal of Solids and Structures, 23(1), 57–93.
19. Buzer, L., (2003). A linear incremental algorithm for naive and standard digital lines and planes recognition. Graphical Models, 65(1–3), 61–76.
20. Cadzow, J. A., (2002). Minimum $\ell 1$, $\ell 2$, and $\ell \infty$ norm approximate solutions to an overdetermined system of linear equations. Digital Signal Processing, 12(4), 524–560.
21. Chan, T. M., (2004). An optimal randomized algorithm for maximum Tukey depth. In: SODA (Vol. 4, pp. 430–436).
22. Chen, D. Z., & Xu, J., (2002). Two-variable linear programming in parallel. Computational Geometry, 21(3), 155–165.
23. Chen, G. H., Chern, M. S., & Lee, R. C. T., (1987). A new systolic architecture for convex hull and half-plane intersection problems. BIT Numerical Mathematics, 27(2), 141–147.
24. Daescu, O., & Palmer, J., (2009). Minimum separation in weighted subdivisions. International Journal of Computational Geometry & Applications, 19(01), 33–57.
25. De, M., Nandy, S. C., & Roy, S., (2015). Prune-and-search with limited workspace. Journal of Computer and System Sciences, 81(2), 398–414.
26. Dobkin, D. P., (1992). Computational geometry and computer graphics. Proceedings of the IEEE, 80(9), 1400–1411.
27. Dyer, M. E., & Frieze, A. M., (1989). A randomized algorithm for fixed-dimensional linear programming. Mathematical Programming, 44(1), 203–212.
28. Dyer, M., Gärtner, B., Megiddo, N., & Welzl, E., (2017). Linear programming. In: Handbook of Discrete and Computational Geometry (pp. 1291–1309). Chapman and Hall/CRC.
29. Edelsbrunner, H., Overmars, M. H., & Seidel, R., (1984). Some methods of computational geometry applied to computer graphics. Computer Vision, Graphics, and Image Processing, 28(1), 92–108.

30. Fan, C., Luo, J., Liu, J., & Xu, Y., (2011). Half-plane Voronoi diagram. In: 2011 Eighth International Symposium on Voronoi Diagrams in Science and Engineering (pp. 127–133). IEEE.
31. Forrest, A. R., (1991). Computational geometry in practice. In: Fundamental Algorithms for Computer Graphics (pp. 707–724). Springer, Berlin, Heidelberg.
32. Goldfarb, D., & Todd, M. J., (1989). Chapter II; linear programming. Handbooks in Operations Research and Management Science, 1, 73–170.
33. Greene, D. H., & Yao, F. F., (1986). Finite-resolution computational geometry. In: 27th Annual Symposium on Foundations of Computer Science (sfcs 1986) (pp. 143–152). IEEE.
34. Gurwitz, C., (1990). Weighted median algorithms for L1 approximation. BIT, 30(2), 301–310.
35. Houle, M. E., Imai, H., Imai, K., Robert, J. M., & Yamamoto, P., (1993). Orthogonal weighted linear L1 and L∞ approximation and applications. Discrete Applied Mathematics, 43(3), 217–232.
36. Imai, H., (1991). Computational geometry and linear programming. In: Algorithms & Architectures: Proceedings of the Second NEC Research Symposium (pp. 9–28).
37. Karp, R. M., (1991). An introduction to randomized algorithms. Discrete Applied Mathematics, 34(1–3), 165–201.
38. Koh, K., Kim, S. J., & Boyd, S., (2007). An interior-point method for large-scale l1-regularized logistic regression. Journal of Machine Learning Research, 8, 1519–1555.
39. Lawler, G. F., Schramm, O., & Werner, W., (2002). Values of Brownian intersection exponents III: Two-sided exponents. In: Annales de L'Institut Henri Poincare (B) Probability and Statistics (Vol. 38, No. 1, pp. 109–123). No longer published by Elsevier.
40. Lawler, G. F., Schramm, O., & Werner, W., (2011). Values of Brownian intersection exponents, I: Half-plane exponents. In: Selected Works of Oded Schramm (pp. 859–895). Springer, New York, NY.
41. Lawler, G. F., Schramm, O., & Werner, W., (2011). Values of Brownian intersection exponents, II: Plane exponents. In: Selected Works of Oded Schramm (pp. 897–930). Springer, New York, NY.
42. Lee, D. T., & Preparata, F. P., (1984). Computational geometry? a survey. IEEE Transactions on Computers, 33(12), 1072–1101.
43. Ludwig, W., (1995). A subexponential randomized algorithm for the simple stochastic game problem. Information and Computation, 117(1), 151–155.

44. Lupu, T., (2016). Loop percolation on discrete half-plane. Electronic Communications in Probability, 21, 1–9.
45. Masajada, J., (2000). Gaussian beams with optical vortex of charge 2-and 3-diffraction by a half-plane and slit. Optica Applicata, 30(2/3), 247–256.
46. Masajada, J., (2000). Half-plane diffraction in the case of Gaussian beams containing an optical vortex. Optics Communications, 175(4–6), 289–294.
47. Matoušek, J., Sharir, M., & Welzl, E., (1996). A subexponential bound for linear programming. Algorithmica, 16(4), 498–516.
48. McCormack, J., & McNamara, R., (2000). Tiled polygon traversal using half-plane edge functions. In: Proceedings of the ACM Siggraph/ Eurographics Workshop on Graphics Hardware (pp. 15–21).
49. McNamara, R., McCormack, J., & Jouppi, N. P., (2000). Prefiltered antialiased lines using half-plane distance functions. In: Proceedings of the ACM SIGGRAPH/EUROGRAPHICS Workshop on Graphics Hardware (pp. 77–85).
50. O'Rourke, J., Chien, C. B., Olson, T., & Naddor, D., (1982). A new linear algorithm for intersecting convex polygons. Computer Graphics and Image Processing, 19(4), 384–391.
51. Onishi, K., & Takayama, N., (1996). Construction of Voronoi diagram on the upper half-plane. IEICE Transactions on Fundamentals of Electronics, Communications and Computer Sciences, 79(4), 533–539.
52. Overmars, M. H., (1988). Computational geometry on a grid an overview. Theoretical Foundations of Computer Graphics and CAD, 167–184.
53. Overmars, M. H., (1996). Designing the computational geometry algorithms library CGAL. In: Workshop on Applied Computational Geometry (pp. 53–58). Springer, Berlin, Heidelberg.
54. Panigrahi, S. C., Alam, M., & Mukhopadhyay, A., (2013). An incremental linear programming based tool for analyzing gene expression data. In: International Conference on Computational Science and its Applications (pp. 48–64). Springer, Berlin, Heidelberg.
55. Pariyani, A., Gupta, A., & Ghosh, P., (2006). Design of heat exchanger networks using randomized algorithm. Computers & Chemical Engineering, 30(6, 7), 1046–1053.
56. Rajagopal, S., (1981). Squeeze casting: A review and update. Journal of Applied Metalworking, 1(4), 3–14.
57. Ray, B. K., & Ray, K. S., (1993). Determination of optimal polygon from digital curve using L1 norm. Pattern Recognition, 26(4), 505–509.

58. Robers, P. D., & Ben-Israel, A., (1969). An interval programming algorithm for discrete linear L1 approximation problems. Journal of Approximation Theory, 2(4), 323–336.
59. Seidel, R., (1990). Linear programming and convex hulls made easy. In: Proceedings of the Sixth Annual Symposium on Computational Geometry (pp. 211–215).
60. Seidel, R., (1991). Small-dimensional linear programming and convex hulls made easy. Discrete & Computational Geometry, 6(3), 423–434.
61. Sethian, J. A., (1999). Level Set Methods and Fast Marching Methods: Evolving Interfaces in Computational Geometry, Fluid Mechanics, Computer Vision, and Materials Science (Vol. 3). Cambridge university press.
62. Smaoui, N., Bérubé, M. A., Fournier, B., & Bissonnette, B., (2004). Influence of specimen geometry, orientation of casting plane, and mode of concrete consolidation on expansion due to ASR. Cement, Concrete and Aggregates, 26(2), 1–13.
63. Späth, H., & Watson, G. A., (1987). On orthogonal linear ℓ1 approximation. Numerische Mathematik, 51(5), 531–543.
64. Thomas, B. G., (2018). Review on modeling and simulation of continuous casting. Steel Research International, 89(1), 1700312.
65. Van, D. B. J., Lee, Y. T., Liu, Y. P., Saranurak, T., Sidford, A., Song, Z., & Wang, D., (2021). Minimum cost flows, MDPs, and ℓ1-regression in nearly linear time for dense instances. In: Proceedings of the 53rd Annual ACM SIGACT Symposium on Theory of Computing (pp. 859–869).
66. Watanabe, I., Watkins, J. H., Nakajima, H., Atsuta, M., & Okabe, T., (1997). Effect of pressure difference on the quality of titanium casting. Journal of Dental Research, 76(3), 773–779.
67. Watson, G. A., (1981). An algorithm for linear L1 approximation of continuous functions. IMA Journal of Numerical Analysis, 1(2), 157–167.
68. Wehr, J., & Woo, J., (1998). Absence of geodesics in first-passage percolation on a half-plane. The Annals of Probability, 26(1), 358–367.
69. Weishan, Z., Shoumei, X., & Baicheng, L., (1997). Study on a CAD/CAE system of die casting. Journal of Materials Processing Technology, 63(1–3), 707–711.
70. Wonisch, A., Polfer, P., Kraft, T., Dellert, A., Heunisch, A., & Roosen, A., (2011). A comprehensive simulation scheme for tape casting: From flow behavior to anisotropy development. Journal of the American Ceramic Society, 94(7), 2053–2060.

71. Yadav, A. K., Kushvah, B. S., & Dolas, U., (2021). Station-keeping error analysis for halo orbits around Libration point L1 using linear control logic. Astronomy and Computing, 35, 100462.
72. Yamamoto, P., Kato, K., Imai, K., & Imai, H., (1988). Algorithms for vertical and orthogonal L1 linear approximation of points. In: Proceedings of the Fourth Annual Symposium on Computational Geometry (pp. 352–361).
73. Yao, F. F., (1990). Computational geometry

CHAPTER

5 Uses of Computational Geometry in Geographic Information Systems

Contents

5.1. INTRODUCTION

Geographical Information Systems (GISs) have acquired significant traction in a wide range of critical application domains during the previous few years (De Floriani et al., 2000). A Geographic Information System is defined by Goodchild, Kemp, and Poiker (1992) as "a hardware system, software, and processes developed to facilitate the management, capture, modeling, analysis, and display of spatially referenced data for the purpose of resolving complicated planning and management issues in a geographic environment" (Figure 5.1).

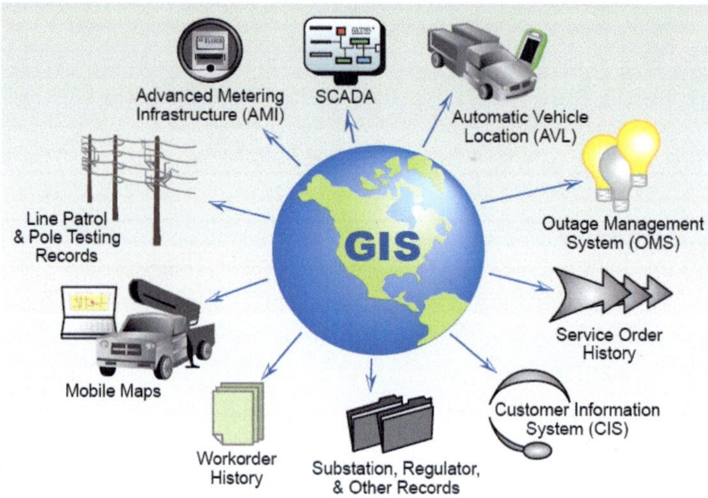

Figure 5.1. GIS has a wide range of applications.

Source: https://jkrbim.com/our-services/geographic-information-system-gis/.

Physical characteristics (size, location, and structure) and non-physical characteristics, which are often recorded in written or numeric format, distinguish geographical data from other types of data. A wide range of challenges are involved in geographic information systems (GIS) as a discipline, including software and hardware equipment for data collecting, standards of data, transmission, and storage, and database administration concerns (like consistency and integrity). This chapter will focus just on difficulties relating to the presentation and processing of geometric components of geographic data, with a great focus on the use of computational geometry approaches to these problems (Figure 5.2) (Smith et al., 1987; Berg et al., 1997).

Figure 5.2. Enterprise GIS system integration and customization.

Source: https://www.ctqconsultants.ca/geospatial/gis-system-integration/.

Classic information retrieval methods had been used to control geographic data in the past, whereas the relevance of geometric features had been frequently overlooked: geometric issues in geographic information system had been treated using ad hoc empirical approaches. These solutions frequently lacked foundations, both geometrically (such as the outcome of an operation described as the output of a specific algorithm) and computationally (such as no complexity analysis, and occasionally had poor actual performance). In geographic information system, the necessity for a strong theoretical foundation and outstanding geometric reasoning capability has become critical. As a result, geographic information system is a crucial subject for the use of computational geometry. Modeling and computational difficulties are involved in the manipulation and representation of geographical elements in a geographic information system. The requirement for algorithms and data structures that offer a suitable compromise among higher computational performance and minimal storage space arises from the necessity to store and analyze large data volumes (Kriegel, et al., 1991; Miller and Wentz, 2003).

The old and novel difficulties in computational geometry and geometric modeling that arise in geographic information system are discussed in this chapter. In a geographic information system, we divide traditional geometric data into two categories: terrain data and map data. Although this categorization is not widely used in the geographic information system field, it is useful for identifying data with varying geographical dimensionality and various types of difficulties. The map data are two-dimensional and placed on the Earth's surface, consisting of points, lines, and polygonal areas that are joined to make subdivisions or arrangements, and are occasionally arranged into layers. Terrain data refers to the three-dimensional arrangement of the Earth's surface. Terrain geometry is represented like a two-dimensional surface, that is, a surface in three-dimensional space characterized by a bivariate function. We also discuss recent research concerns in geographic information system relating to the handling of completely three-dimensional data, furthermore these two types of traditional geographic information system data (Mark, 2003).

5.2. MAP DATA MODELING

5.2.1. Two-Dimensional Spatial Entities and Relations

Lines, points, and regions contained in a 2D space have been termed two-dimensional elements in the perspective of geographic data processing. Smaller areas of the Earth have been frequently projected onto a flat surface and modeled utilizing traditional Cartesian aligns. Azimuthal projections, which utilize a plane curve to the Earth in the center of the area to be cylindrical projections, conical projections, and mapped, which utilize a cylinder curve to the Earth alongside the Equator or alongside the line of longitude, have all been examples of projections (Transverse Mercator projection; Figure 5.3) (McKeown, 1987; Overmars, 1996).

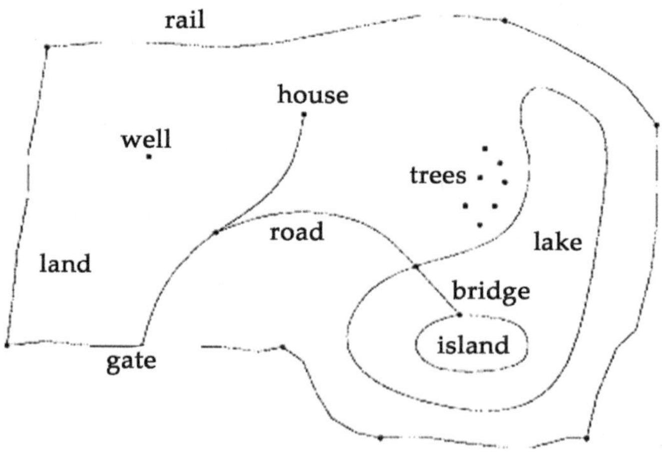

Figure 5.3. A 2D example of a geographic map.

Source: https://www.sciencedirect.com/topics/computer-science/spatial-entity.

The definition of a global reference system is a much more difficult task to solve. For this reason, a variety of conventions have been employed. There are several approaches, including subdividing the Earth's surface into 'x' zones and projecting every zone differently (the Universal Transverse Mercator utilizes 60 various areas); other methodologies include the utilize of a spherical direct system dependent upon long-latitude coordinates or the projection onto an ellipsoid (Shekhar et al., 1998; Held and Williamson, 2004). Furthermore, in the sequel, we would always suppose that the map data is contained on a Euclidean plane, in accordance with the local convention. The entities used in GIS have been broader than those used in computational geometry (Figure 5.4).

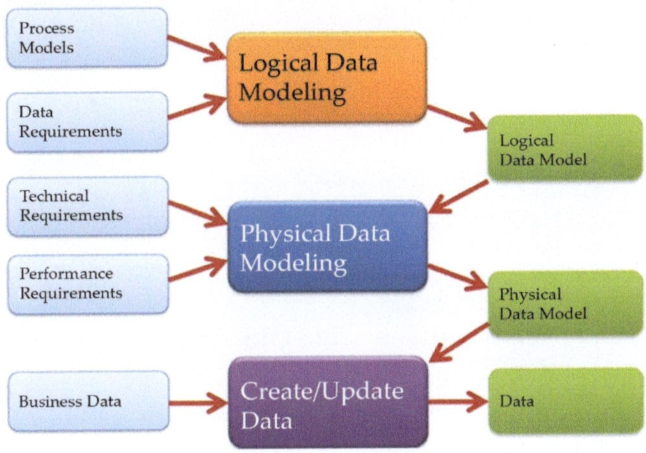

Figure 5.4. Forms of data model.

Source: https://www.guru99.com/data-modeling-conceptual-logical.html.

In certain situations, lines cannot be simple; regions cannot be simply linked (such as they may encompass holes' 'j' and internal Faeroes, such as dangling lines (cuts), and isolated points). A two-dimensional instance of a geographic map is shown in Figure 5.3. The existence of gaps and characteristics in algorithms and data structures for processing geographic entities may have an impact on their design. The majority of geographic computational algorithms primarily work with piecewise-linear geometric elements. Even though there have been data structures and strategies for expressing and manipulating curved things, this chapter does not cover them.

Spatial relationships are based on the relative locations of spatial items and are established on pairs of them. The description, categorization, assessment, and research of invariants of spatial connections has a thriving literature (Samet, 1990; Fisher and Unwin, 2005). Despite this, a homogenous and comprehensive characterization (for example, dependent upon an algebraic technique) remains elusive. It's fairly unusual to come into real situations and applications for which none of the models given in the writing are adequate. Spatial relations, on the other hand, are commonly characterized as given below:

- Incidence, containment, adjacency, overlapping, and other topological relations have been defined in words of geometric intersections among borders, interiors, and complements of things (Câmara et al., 1996; Wassan, 2015);
- A metric is specified in words of distance among items, such as closest neighbors or range queries;
- Directional relations are defined by the relative positions of things in certain oriented direction: instances include before, after, left-of, right-of, south, north, etc.;
- Order relations are characterized through qualities that cause partial order to be induced on sets of things. Containment provides an instance (Batra and Marakas, 1995; Wise, 2000).
- Since many activities in GIS require the assessment of geographical connections among a query object and a set of spatial entities, longitudinal links play a key role.

  The following is a wide taxonomy of geographical datasets that is useful for both modeling and calculation
- Assemblages of spatial things (usually arranged into multiple layers) that may occupy any relative location on the plane and intersect at will;
- We discuss how to divide a compact planar domain into a set of non-overlapping elements. Collections can define arrangements that have been more generic than those explored in computational geometry because they may incorporate irregular things of every shape and size. Subdivisions, in a similar manner, have been more generic than the subdivisions utilized in computational geometry because they can comprise areas with characteristics that are related to one another many times. Subdivisions have been more straightforward than collections. The combinatorial difficulty of a subdivision is proportional to the total number of entities in the subdivision, but the arrangement suggested by a collection

can have quadratic combinatorial complexity. The topological relationship between two things in a compilation can be any feasible topological relationship, but the topological relationships between two entities in a subdivision can only be adjacency and incidence between its component entities, as well as the presence of characteristics (Batra and Davis, 1992; Herrera et al., 2009).

5.2.2. Raster and Vector Models

In the geographic information system field, there has long been a discussion about whether it has been preferable to represent geographic data using a raster and vector model. Throughout all tasks, neither model is considered to be better. "A typical approach is to distinguish among an entity-based perspective, wherein space is formed from things that occupy space, and a space-tailored perspective, in which every point in space has certain qualities," Frank (1988) explains. This is a philosophically well-established viewpoint." These two distinct interpretations of spatial conceptions appear to correlate to the raster and vector techniques.

Spatial entities are represented visually in a vector model by their shape and properties. Subdivisions and collections may both be displayed, and topological relationships between things can be explicitly marked. An explicit depiction may support a degree of precision that is arbitrarily high. Because the vector model has been entity-based, it's ideal for searching a spatial database with spatial entities as search keys. Effective space-oriented access, on either side, may need advanced search structures and strategies. Spaghetti models, or initial depictions in vector data modeling, comprise collections of unconnected basic polygonal lines and points. Regions are not explicitly encoded in a spaghetti model, but they are tacitly characterized by their borders. Because the spaghetti standard is essentially a collection of things with no recorded relation, it can encode either a subdivision or a collection. The disadvantage is that there is practically no assistance for effective processing of query. In the subsections that follow, highly efficient information structures have been described separately (Figure 5.5) (Fullea et al., 2015).

The domain of a raster model has been segmented into a huge number of atomic sections, akin to picture element in a digital image. Every pixel contains information about the area it encompasses. The pixels in which points are located are used to identify them. Regions and lines are obtained as pixel aggregations (digital regions and lines) based on membership qualities, whereas regions and lines have been obtained as pixel aggregations (digital regions and lines).

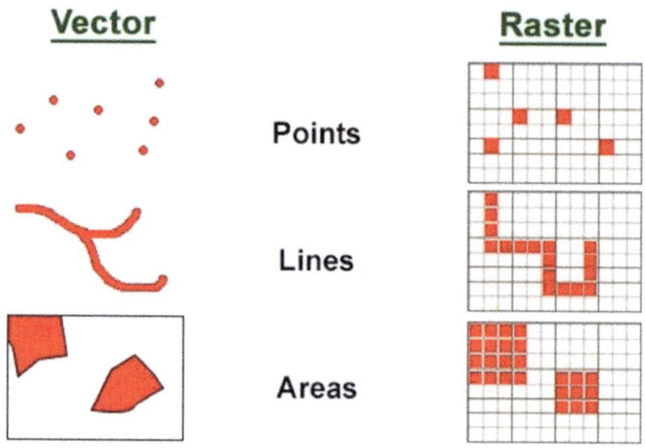

Figure 5.5. Raster and vector models are depicted.

Source: https://in.pinterest.com/pin/data-structures-data-raster--466615211373107260/.

Special elements have been thus not directly characterized in a raster, although they may be derived via relations which group pixels based on their shared attributes. The raster model describes a domain subdivision via default; however, a collection may be created by combining numerous members per pixels (Worboys et al., 1990; Arayici, 2007).

5.3. MAP DATA PROCESSING

5.3.1. Spatial Queries

Geometric information, such as the location and size of the items in question, is all that is needed to answer a spatial query. It is possible for a query object to be either part of the spatial database or not, and for a query space 'S' to be just the entire spatial database or only a subset of it retrieved using appropriate filters, as well as for a spatial relation 'R' to be either part of the database or not. The following is an example of a generic query:

'R' with 'q' returns all objects that have been in a relation.

The categorization of spatial questions follows the categorization of spatial relations in a direct line (Figure 5.6).

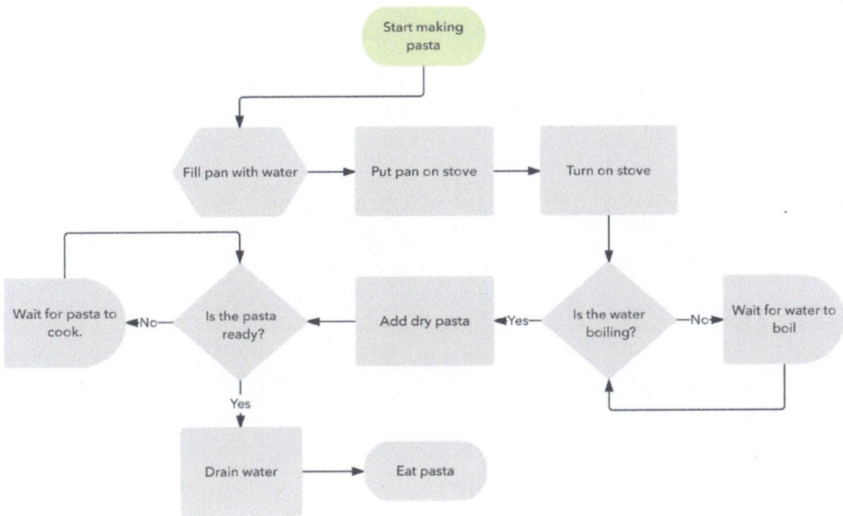

Figure 5.6. Flowchart for process mapping.

Source: https://www.lucidchart.com/pages/process-mapping.

The nature of the request space (as well as the model that represents it), the nature of the inquiry object, as well as the nature of the relative all influence the calculation strategies that may be used to answer spatial questions. Furthermore, a simple comprehensive examination of all options reveals that most spatial inquiries may be simplified to several basic computational geometry issues (Mendonça-Santos et al., 2008; Costa et al., 2017).

Research on the validation and solution of metric and topological requests on geographic maps represented using a cell complicated is described in (Livingstone et al., 1997). When the question object belongs to the map, topological questions are handled by map traverse, but when the query entity does not belong to the map, they are always reduced to polygon intersection, line intersection, or point location.

In Lomov et al. (2007); Fonseca and Gaspar (2021), an effective technique for map traversal in geographic information system was presented, which offers the foundation for the answer of numerous topological questions on a map. Without using mark bits in the stack or a data structure, this technique traverses a planar section and reports all of its areas. As a result, it's particularly well suited to a map kept in a read-only geographic information system database (Figure 5.7).

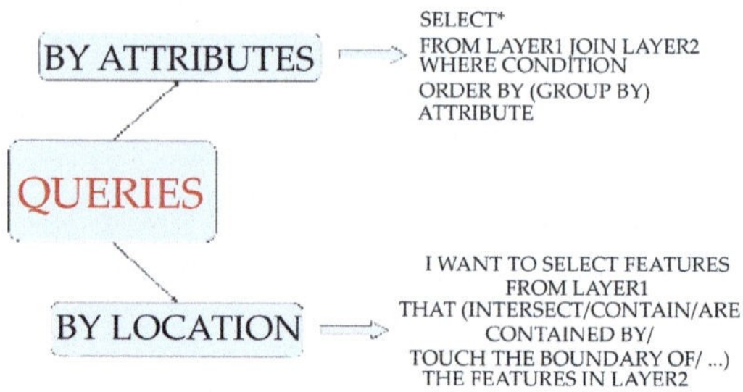

Figure 5.7. Kinds of spatial analysis queries.

Source: https://www.researchgate.net/figure/Query-types-in-spatial-analysis_fig4_279472747.

Metric queries, on the other hand, have not completely backed by the map's topologic structure and must rely on auxiliary structures like spatial indices or Voronoi diagrams to be effectively replied. Gold and Roos (1994) have studied the usage of Voronoi diagrams (Bornberg-Bauer and Paton, 2002; Sousa and Hey, 2013) for addressing geographic queries in geographic information systems. The Voronoi diagram can be thought of as a bridge among the vector and raster models. "The Voronoi model has the whole tiling feature of the raster model," Gold (1994) observes, "but the tiles have been dependent upon the proximate region around every object." In the computational geometry literature, both dynamic and static methods for building a Voronoi diagram of 0-, 1-, and 2D items in the plane are devised. The Voroni diagram, as demonstrated by Gold and Roos, is a constant, locally editable data structure suited for algorithms such as polygon labeling, buffer zone creation, navigation queries, map intersection with a line segment, surface interruption at a novel location (Amadi et al., 2002), and so on.

Collections on queries of entities may advantage from less explicit help from data structures' relations. In theory, computational geometry systems that work on procedures can provide answers to certain of these questions, but they have been seldom used in reality, possibly since they have been difficult to develop and have a higher computational complexity.

Spatial indexes, which have been simple to construct and operate well in the typical scenario, are preferred strategies for processing collections of items. Wang et al. (2000); Emsley et al. (2002) presents a detailed investigation of the average-case difficulty of accessing quadtrees. Various techniques for spatial inquiries dependent upon spatial indexes are provided in Milton and Kazmierczak (2004); Allen and Tibshirani (2012).

Non-geometric properties of objects may be coupled utilizing spatial inquiries. The careful study in this scenario should consider both non-geometric and geometric factors. The creation of appropriate integrated models and procedures, and also ways for assessing their performances, are hot open challenges that demand multidisciplinary study.

5.3.2. Map Overlay

Information is organized into layers (for example, usage of land, hydrography, road network, etc.), in GIS and maps are created by overlapping layers of interest. The edge method provides the foundation for resolving complicated spatial problems. The input maps have not been only overlapped, however also merged depending on their properties in these searches (for example, through utilizing Boolean expressions). Just the geometric part of map overlay is discussed herein, wherein computational geometry methods may be utilized (Figure 5.8).

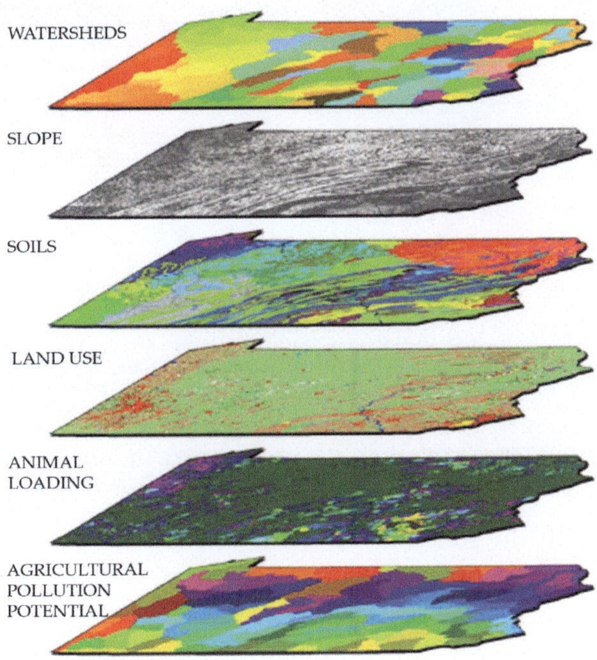

Figure 5.8. An illustration of a map-overlay.

Source: https://www.e-education.psu.edu/natureofgeoinfo/c9_p6.html.

Overlaying raster maps is as simple as merging the attribute values of comparable pixels and does not need the use of computational geometry algorithms. To match raster grids with varied orientations or cell sizes, conventional modifications have been utilized (Hadzilacos and Tryfona, 1996; Papy et al., 2007). The overlaying of vector maps is the topic of this article.

There have been various types of overlays in geographic information systems, including overlay among two subdivisions. Point-to-area overlays and 'l' Line-to-area intersections are two types of overlays. Furthermore, input information for both line-to-area and area-to-area intersections can be sets of potentially overlapping lines and regions. Two maps can be merged to intersect, create a union, or update each other utilizing Boolean expressions; many maps can be merged utilizing Boolean expressions (Figure 5.9).

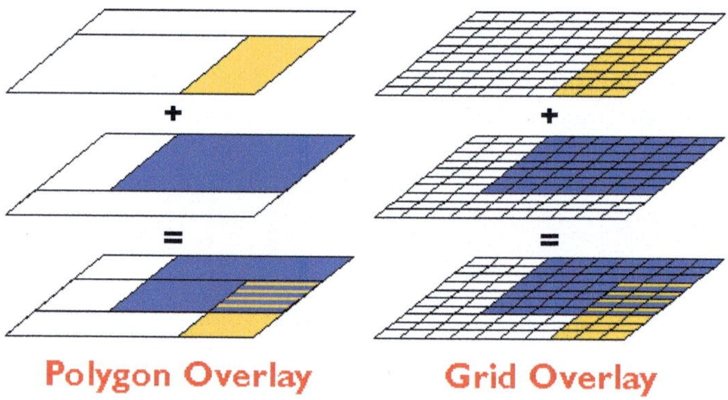

Figure 5.9. Overlay of polygons and grids as an instance.

Source: https://www.e-education.psu.edu/natureofgeoinfo/book/export/html/1771.

Overlay issues are treated in a very typical manner in computational geometry:

- Find all the junction locations inside an arbitrary collection of segments in the segment intersection issue;
- Find the junction sites of two sets of sections when no intersections exist inside the same set in the red-blue intersection issue;
- The putting plane subdivisions on top of one other. Calculate a sub-division Z' whose regions and edges have been derived as the junction of S_1 and S_2's regions and edges, given two subdivisions £1 and Z2.

Line maps that overlap may be created using segment intersection techniques (such as road networks). The methods in the colors blue and red are applied to map layers that contain broad subdivisions containing attributes. There are further limits on certain plane subdivision methods. For example, certain are restricted to trapezoidal maps (Bonham-Carter and Bonham-Carter, 1994; Azab, 2014), while others are restricted to convex areas (for example, a triangulated map) (Barreira-González et al., 2015; Alrashed et al., 2018).

The k fi(n^2) has been the worst-case number of intersections across all overlay situations wherein 'n' is the input size, and 'k' is the maximum number of intersections. The complexity of the task is limited to O (n + k). In a brute force solution, the intersections among each pair of objects from the two supplied layers are computed, and the solution is effective in é9(n^2) layers (Figure 5.10).

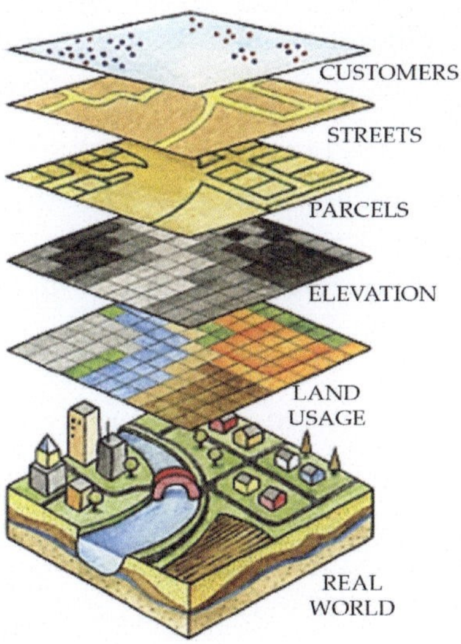

Figure 5.10. Overlay of a map showing land usage.

Source: https://saylordotorg.github.io/text_essentials-of-geographic-information-systems/s11-02-multiple-layer-analysis.html.

In the geographic information system community, the majority of techniques are heuristic "filters," that attempt to decrease the number of tested pairings via superimposing a classified spatial index on top of the data. Filtering is accomplished by breaking the plane into blocks and dividing the objects to be intersected (the regions and edges of every layer) across the different blocks in a predetermined pattern. Due to the fact that they may not collide, pairs of objects that are located in separate blocks of space may be removed from consideration. Furthermore, in every block, an extensive search for intersections has been performed using the results of the previous study. The filter, on the other hand, is not assured to be successful, which means that certain duplicate tests may still be performed. Such approaches, on the other hand, have excellent practical behavior. Although if they result in an extensive intersection study in the worst-case scenario, its performance is extensively tested in the real world through extensive experimentation. Some approaches (e.g.: Holroyd and Bell, 1992; Cromley and Hanink, 1999) provide a theoretical analysis of the average randomized instance that may be used to validate the method.

There are two types of filtering methods:

- The methods that decompose the space into disjointed blocks, eventually dividing items that span many blocks; and
- The methods that decompose the space into potentially overlapping blocks, preserving every object in precisely single block.

A uniform grid (Maffini, 1987; Brisaboa et al., 2017) has been the basic decomposition approach dependent upon discontinuous blocks that has been proposed. The resolution of the grid is what affects the efficacy of the filtering system in question. When data is spread unevenly, homogeneous grids perform poorly, and this is especially true when the data is large.

Quadtrees (da Paz et al., 2008; Taillandier et al., 2016) are adaptable grids that are formed by dividing a basic square into four sectors in a recursive manner (Albert and Golledge, 1999; Jenness, 2004). As a result, depending on the density of data in each section of the domain, a different grid resolution may be reached in every portion of the domain. Several other quadtree-like structures are proposed, each with its own set of advantages and disadvantages based upon the kind of data to be stored: either as the edges or the regions of a map. Boolean operations have been simple to perform on maps that have a quadtree-like structure placed on them. Each of the two input quadtrees has been explored synchronously, with visits to nodes relating to a similar piece of space occurring at a similar time; a quadtree representing the outcome of the Boolean function has also been constructed progressively. Binary space partitioning (BSP) is an image processing approach that has recently been used to GIS (Battersby et al.,

2006). It employs a more flexible scheme that is nevertheless based on a cyclical subdivision of space. At each step, an imaginary straight line divides the current block into two pieces. The resulting structure is a binary tree with nodes that relate to curved polygons. Algorithms for conducting Boolean functions on objects stored in BSP trees are supported (Li et al., 1999).

By comparing bounding boxes, techniques rely on contiguous blocks to filter candidate pairs. A 2D entity's bounding box is the smallest axis-parallel rectangle that entirely encompasses it.

The R-tree family of structures (Newcomer and Szajgin, 1984; Veregin, 1989) is made up of trees with a degree bound by a constant number, where data items are placed at the leaves as well as every interior node symbolizes the bounding box of all the objects contained in its offspring. Distinct bounding boxes for various nodes may overlap. For each map layer, a distinct R-tree is created. While descending the trees, junctions between layers are computed: tests on the bounding boxes are done at each step, and the results are utilized to chop out some branches of the trees that are assured to create no junction. According to intuition, an "ideal" R-tree should be symmetrical, its bounding boxes should be as disjointed as feasible, and the quantity of area covered inside them should be as small as possible. The filter's efficiency is improved as a result of these characteristics. Among the several R-tree varieties currently in existence are those that are distinguished by distinct criteria for determining "excellent quality" in an R-tree, as well as by different construction methodologies. A number of common criteria include attempting to reduce the area covered by the rectangles | (Egenhofer and Richards, 1993), avoiding overlapping rectangles (van Oosterom, 1994), and grouping objects centered on their spatial proximity (Kriegel et al., 1991; Kuitert, 2013). Because several R-trees allow for dynamic item deletions and insertions, there is a trade-off between the amount of time it takes to update the tree and the quality of the tree that is created. In practice, R*-trees] appear to be the most effective of the trees.

Computational geometry has been studying algorithms with low theoretical complexity (Arbia et al., 1998; Wickens et al., 2000) with the goal of finding efficient methods. Several of these techniques, particularly the plane sweep paradigm, have been combined with GIS environments. The classification of the major existing approaches can be as follows:

- A method created on a segment tree, which is a specific data structure that hierarchically separates the plane based on the arrangement of input segments (Huang et al., 2017);

- A randomized incremental technique, in which all the crossings formed by a general set of segments are computed one at a time by adding the sections one by one;
- A sweep-line method for finding crossings by moving a line along the plane and observing the sections crossed by that line (Herbei et al., 2011; Verstraete, 2015);
- An oR topological sweep centered on an algorithm, i.e., sweeping driven by the hierarchical model of the stored subdivisions.

Herbei et al. (2011) addresses the issue of overlapping subdivisions encoded by hierarchical standards. A segment tree is a binary tree whose leaves reflect the vertical stripes defined by every pair of successive vertical lines drawn through into the segment endpoints, in order from left to right. The data segments are correctly linked to the tree's nodes. Then it's just a matter of testing intersections between segments belonging to the same node. By dynamically generating just a section of the tree at a time, such as a single level (Shahzad et al., 2020) or a single root-to-leaf path (Richards and Egenhofer, 1995; Gaeuman et al., 2005), space requirements are decreased.

Segment intersections are provided in incremental algorithms when creating a triangular decomposition of the plane. Drawing a vertical line across each section endpoint or intersecting point till another portion of 'S' is found in both directions yields the trapezoidal decomposition generated by a set 'S' of sections. Unique auxiliary data structures that encode the construction history improve the efficiency of locating the trapezoid bone of the present decomposition impacted by the introduction of a recent segment (Table 5.1).

Table 5.1. Major Map Overlay Algorithms

Algorithm	Approach	Input	Time
1	Incremental	Segments	O(n log ii k)
2	Segment tree	Red-blue segments	O(ii log n + k)
3	Sweep-line	Segments	O((n I) log n)
4	Sweep-line	Plans subdivisions with curved regions	O(n log n k)
5	Sweep-line	Red-blue segments	O(n log n + k)

6	Sweep-Zinc	Portions	O(n log n – I – k j)
7	Topological sweep	Plane subdivisions	O(n+k)

Note: The worst-case runtime and input requirements are listed for each algorithm. The input and output sizes are denoted by the parameters 'n' and 'k,' respectively; the time complexity of incremental algorithms has been calculated using a randomized method (Figure 5.11).

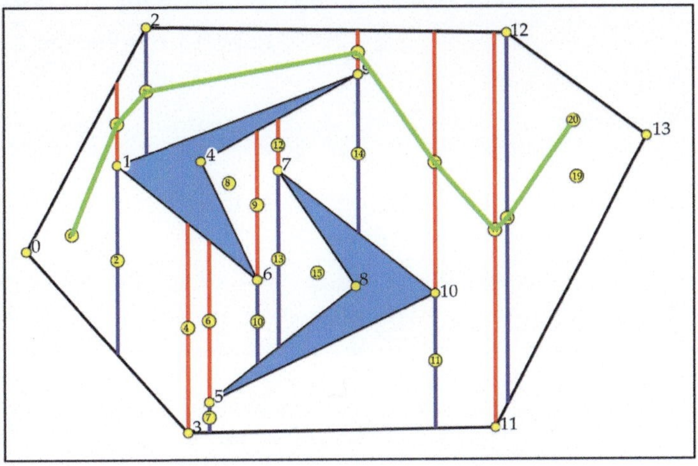

Figure 5.11. Upper and lower edge extensions of a polygon vertex.

Source: https://user.ceng.metu.edu.tr/~akifakkus/courses/ceng786/hw3.html.

5.3.3. Map Generalization

In cartography, map generalization refers to the process of selecting and representing particulars to be exhibited on a map at a specific scale. Various activities are involved in map generalization, such as deleting items, reducing geometry, aggregating multiple objects into one, and so on. In the context of GIS, generalization refers to the process of reducing "information density" in a geographic file while maintaining its whole structure and interpretation. Both conceptually and computationally, generalization is a difficult task. It includes non-geometric tones as well as geometric characteristics of geographic entities (form, structure, and detail). Generalization procedures are classified into the following basic operations in traditional cartography (Magalhães et al., 2015; Song et al., 2019).

In the simplified map, selection determines whether an entity should exist or not:

- Aggregation is the process of combining multiple elements into a single entity;
- The dimension of an entity is reduced when it is collapsed;
- Symbolization is the process of converting a geometric item into a symbol;
- Simplicity reduces the precision of portraying an entity's shape;
- Exaggeration alters an entity's shape to emphasize aspects that would otherwise be invisible on a tiny scale;
- Movement also alters the shape, separating entities that might otherwise be too close together.

Non-geometric rules are typically used to govern the selection and aggregation processes. Symbolization is also non-geometric, with the exception of the challenge of avoiding symbols from overlapping. The subsequent operations are mostly geometric in nature, as they have an impact on the geometry of entities that are explicitly shown on the map at lower scale (Figure 5.12).

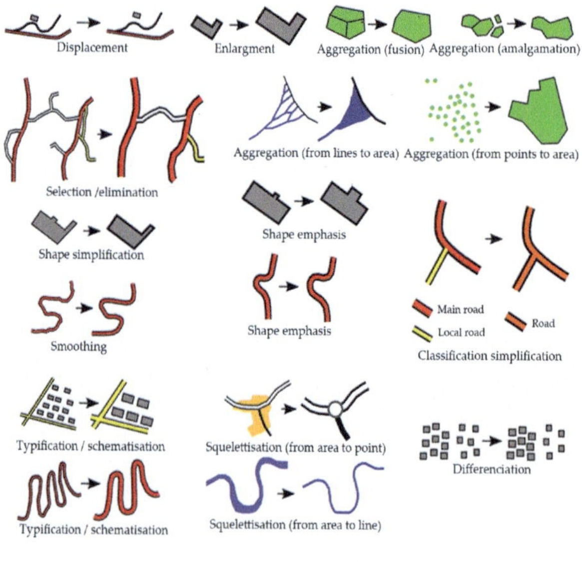

Figure 5.12. Some map generalization operations.

Source: https://www.researchgate.net/figure/Some-generalization-operations-translated-and-adapted-from-Musti-re-2001_fig2_220144000.

The most researched problem in generalization is simplification. The most well-known and widely used approach is arguably Douglas and Peucker's early refining heuristic for a basic open triangular chain (Wu et al., 1990). Starting with an early estimating order, the method additions the point that maximizes the distance from the present calculation at each cycle, because the greatest error stays below a predetermined limit. In the worst situation, a simple implementation takes O(n') time; a more sophisticated approach that takes O(n log* n) time has just been presented (Buttenfield and McMaster, 1991). Other heuristics for the same goal have been given in the GIS literature, and they are typically empirically contrasted (Figure 5.13).

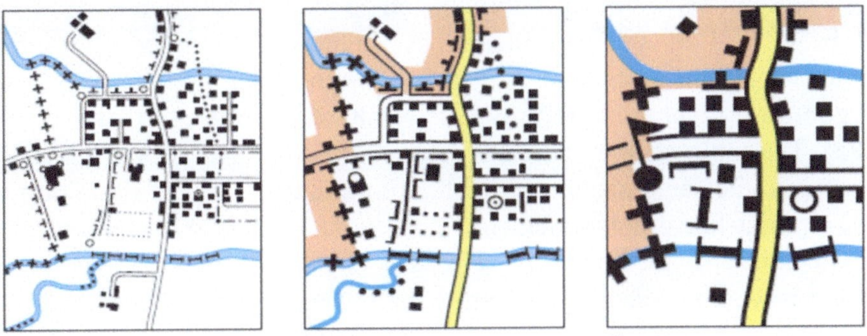

Figure 5.13. Automated map generalization.

Source: https://infovis-wiki.net/wiki/Teaching:TUW_-_UE_InfoVis_WS_2007/08_-_Gruppe_05_-_Aufgabe_1_-_Generalization.

Imai and Iri (2016) are responsible for the first formalization of line reduction as an optimization problem, which allowed for a more rigorous approach to the issue. It is specifically stated that, given an input slope and some metric to measure the error in estimating it, they plan two fundamental problems: (1) minimization of output chain vertex count for a given threshold error; and (2) minimization of output chain error for an input curve as well as some metric to measure the error in approximating it. Imai and Iri provide algorithms for both problems, as well as a list of findings from other authors, for each problem using four different metrics. The authors Guibas and Stolfi (1985) also mention some other findings. For generic polygonal chains, all algorithms have a super-quadratic time complexity, however, for a particular metric, a super-quadratic time complexity can be reached (Jones et al., 1995;

Li, 2007). The trouble I can be solved in linear time in the specific case in which the input is the image of various points | 148 J. Problem I can be solved in linear time in the case in which the input is the image of various points | 148 J. Guibas and colleagues (1985) also investigate Problem I as a problem of ordered stabbing. Only discs that are centered at the vertices of the input chain are taken into consideration; the solution must be a minimum-link chain that stabbings such discs in the proper order. They provide an O(n log n) suboptimal greedy method for such a problem, in addition to an optimal quadratic solution based on dynamic programming: Completely, they provide a linear time-optimal solution for the exceptional scenario in which the discs in the input are all disjoint in the linear time domain.

Because none of the algorithms listed above can ensure that the output will be devoid of self-crossings, they cannot be regarded as ultimate solutions to these problems. The task of finding a minimal-level simple polygon for a given homotopy type is NP-hard, according to Guibas et al. (1985). The topic of line generalization is also examined in the framework of a map in the same paper: establish a minimum-link approximation for a plane straight-line graph generated by a set of non-crossing polygon chains given a plane straight-line graph. This task is also NP-hard: the challenging element is locating vertices with more than two interacting edges.

Weibel (1988); and de Berg et al. (2000) have lately addressed a more general topic of subdivision simplification. Weibel (1988) specifies simplification limitations based on cartographic principles, which are divided into metric, topological, semantic (such as class membership preservation), and gestalt principles. De Berg et al. (2000) tackle the problem of reduction with the goal of minimizing a subdivision's complexity. For the problematic, they offer a nearly quadratic algorithm.

Merging areal features has gotten a lot of attention, although there hasn't been much research done on it. Earlier methods were often raster-based and used mathematical sound structure operators like erosion and dilation. The medial axis transform is used to solve collapse problems such as reducing a ribbon-shaped area to a centerline or a spatial feature to a point. Jones et al. (2004) demonstrate how to implement a few generalization operators using a triangulated spatial model of a map based on limited 3D shape of the map edges and vertices (see Ware and Jones, 1998; Regnauld, 2001). Conflict resolution and detection, dimensionality reduction via skeleton generation, border interpretation, and integration of close objects are all possible with these operations (Figure 5.14).

Figure 5.14. Scaling and generalization of google map.

Source: https://www.axismaps.com/guide/scale-and-generalization.

On a broader level, Saalfeld (1974) newly proposed that the fundamental processes of map generalization can be defined in terms of graph drawing. Even if such an approach appears to be encouraging, it is now only a challenge for graph drawing professionals, with no algorithms based on it presented. Puppo and Dettori have recently presented a new approach to map generalization based on rules generated between cell complexes and constant un-der the finite topology. This method is mostly used to provide consistent control over actions that change the map's topology and to define multiresolution models. Del Los and Hadzilactis have recently suggested geometric consistency algorithms based on the methodology of Puppo and Dettori.

5.4. TERRAIN DATA MODELING AND PROCESSING

Terrain handling necessitates different input, data models, and techniques than 2D data processing. Terrain processing is also a key source of geometric difficulties, which encourages the usage of computational geometry in both employing and inventing new solutions.

A model depicting terrain relief depending on a finite number of samples is referred to as a digital terrain model (DTM) (Stoter et al., 2009; Fang et al., 2017). Elevation data is collected via sampling techniques or by digitizing existing contour maps (Figure 5.15) (Câmara et al., 1996).

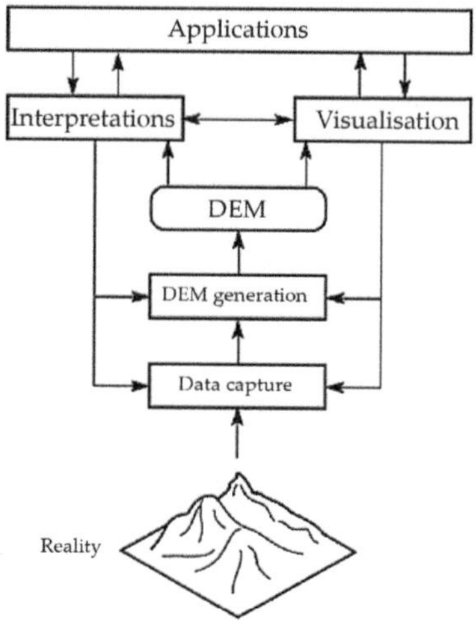

Figure 5.15. The main tasks of the digital terrain model.

Source: https://www.researchgate.net/publication/320024823_Creating_and_analyzing_the_Digital_Terrain_Model_of_the_Slivovo_area_using_QGIS_software.

Different types of models can be constructed from sample data, including contour maps, rosters, and mathematical surface models. Mathematical models can be either global or local. Regularly distributed data leads to regular square grids (RSGs) in most cases, but randomly distributed data leads to triangulated irregular networks in almost all cases. Given that we will be focusing on models and methods that are rooted in discrete mathematics, we will not cover raster models and the digital geometry techniques that are used to process them, nor will we cover global mathematical models or algebraic procedures. The remainder of this section will deal primarily with general issues relating to the storage and construction of DTMs, with specifically issues relating to their analysis being discussed in the following section, and so on. If you are interested in more technical concerns about DTMs, and especially algorithms on TINs, you may refer to a study by van Kreveld (1997), and you can refer to a survey by Magillo and Puppo (Floriani et al., 2000) for difficulties on parallel algorithms for terrain analysis.

5.4.1. Classical Terrain Models

- A DTM (digital terrain model) is a model based on a finite collection of data that provides information on such a landscape.

Terrain data include elevation measurements at a series of coordinates V $[v_0, \ldots, v_N]$ C, as well as perhaps a set of non-crossing straight-line segments $E = e_o, \ldots, e_m$ with ends in V. The data points in 'V' can be dispersed or create a pattern.

- A standard grid lines in £ can be used to create polygonal chains. In the context of GIS, three types of DTMs are commonly studied (Peuquet, 1984; Arayici, 2007).
- Polyhedral terrains are the representation of a piecewise-linear function. The division of the domain f_2 into polygonal regions with vertices in 'P' can be used to describe a polyhedral terrain model. A planar patch is the image of "f" over each region. Triangulated irregular networks (TINs) are the most often used polyhedral terrain models, in which all regions are triangles.
- A gridded elevation model is known as a domain partition into polyhedra produced by a regular grid over H. The RSG, in which all areas are squares, is the most often used gridded model. Function 'f' is often a bilinear function that interpolates the grid's vertices. An RSG may be regarded as a raster in some circumstances, with a constant function applied to every region. This is known as a stepped model since it is non-continuous around the grid's boundaries.
- Contour maps – a digital contour map is an estimate of a set of contours given a series 'Q' $(q_0, \ldots, q_h]$ of real values. Contours are frequently represented as point sequences in digital contour maps; a line interpolating points of a contour can be created in a variety of methods, ranging from the simplest instance of a polygonal chain through spline curls of various orders (Figure 5.16).

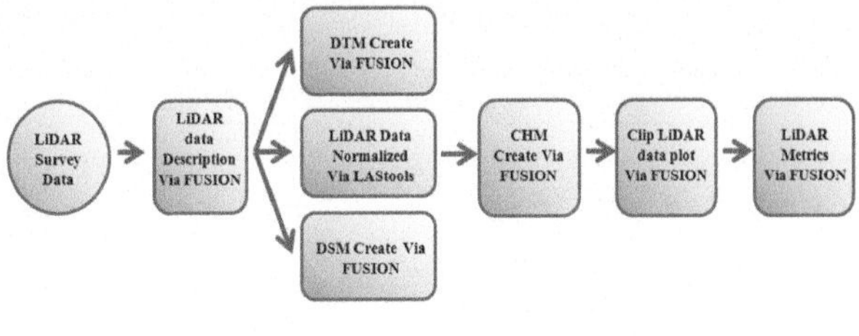

Figure 5.16. Steps to individual tree processing.

Source: https://www.researchgate.net/figure/Steps-to-individual-tree-processing-DTM-Digital-Terrain-Model-DSM-Digital-Surface_fig2_236620622.

In the context of scattered data interpolation, optimality conditions specified on the estimating surface have also been explored. Triangulations specified by these criteria are called data-dependent triangulation as the optimization is based on the z-values of data points as well as their x-v aligns. Dyn, Rippa, and Levin (1990) look at optimality criteria that may be expressed as a cost function on the resulting surface. The following are some examples of optimality criteria:

- Criterion with three dimensions – on edges and triangles of the surface, instead of their projections on the plane, requirements such as the min-max angle, the max-min angle, and the minimal edge length are evaluated;
- Almost C1 criterion – on the edges of the triangulation, the piecewise-linear surface that is as close to a C' approximation as possible is chosen;
- The unidentified surface is assumed to be the one that minimizes some function specified across an appropriate set of functions, according to variational criteria. The roughness, which is determined by the function's Sobolev semi-norm, is an example of such a function. Surprisingly, regardless of the c-values tf its vertices |2.3flJ, the surface with the least irregularity is always supplied by the 3D shape of projected points.

5.4.2. Terrain Generalization

The more data there is, the more accurately a terrain may be represented. Modern acquisition methods produce large datasets that allow for realistic terrain modeling. Nevertheless, a high level of representation accuracy comes at a cost in terms of storage and processing. By balancing representation costs and accuracy, generalization approaches can bring such costs into reasonable boundaries (Figure 5.17).

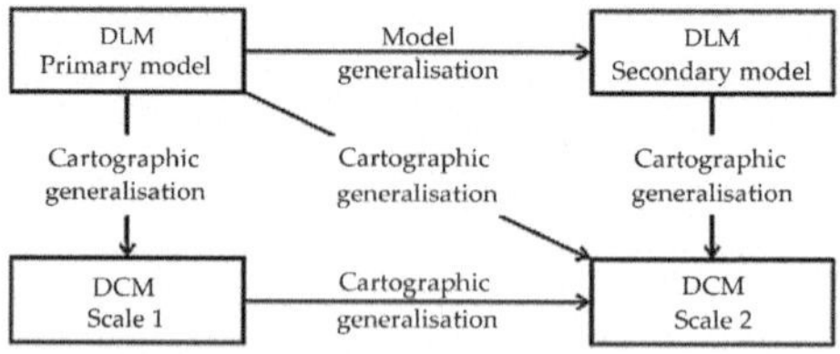

Figure 5.17. Terrain generalization flowchart.

Source: https://link.springer.com/chapter/10.1007/978-3-319-00203-3_8.

A reduced set of data is used to create an approximate terrain model. The approximation error ‘s’ is calculated using a reference model constructed from the entire data set: a general choice is the maximum difference between the height at a data point and the model’s inserted elevation. The inverse of the error, namely, 1/, can be used to determine the accuracy of an estimated depiction to the reference model, which has no relation to the precision of samples to reality (1 – 1 – c). An application’s performance can be improved by using an approximation terrain model with a required level of accuracy. The edge may be changeable over the surface domain in some circumstances. For terrain analysis, for example, better precision may be required near-critical features; for landscape visualization, the accuracy may decrease with distance from a particular perspective. However, by conducting a simple rescaling of the input data using the inverse of the threshold function (Chauve et al., 2007; Trantham and Kennelly, 2022), a generalization problem with a variable threshold can be easily transformed into a problem with a fixed threshold. The desired goal of terrain generalization is to achieve the best possible balance of accuracy and representation size (Figure 5.18). There are two alternative optimization criteria for line simplification:

- for a given accuracy, reducing the number of vertices in the model; and
- for a given number of vertices, optimizing accuracy.

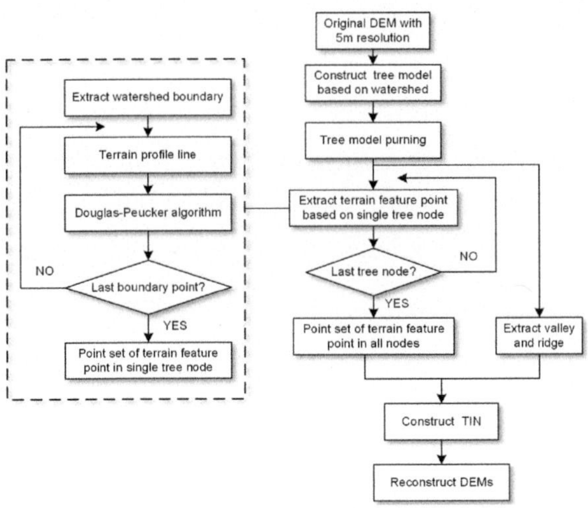

Figure 5.18. Generalization method based on watershed and tree structure.

Source: https://journals.plos.org/plosone/article/figure?id=10.1371/journal.pone.0159798.g002.

When it comes to the first problem, Agarwal and Sun (1998) have demonstrated a negative result. They take the case of a polyhedral terrain and show that the issue of estimating it at a benchmark with another TIN having the smallest number of vertices and triangles at any location is NP-hard. As a result, it follows that the 2nd challenge is also NP-hard. When vertices of the estimated terrain are confined to lie at new data points, it was hypothesized that such problems would still be NP-complete (see Figure 5.18).

The first problem is the only one for which there are algorithms that can produce a find a good in polynomial time while still ensuring some limitations on the solution's size. Using a first technique proposed by Agarwal and Suri (1998), it is possible to construct an approximation solution in O(n) time with O(k log k) triangles, where 'k' is the number of triangles in the best solution. The algorithm is described in detail in the paper. Agarwal and Desikan (2000) proposed a randomized algorithm that can achieve a size of $0(k_2 \log 2\ k)$ in $0(n_2 + h\ k_3 \log 2\ k \log n/k)$ anticipated time and can achieve a size of $0(k_2 \log 2\ k)$ in $0(n_2 + h\ k_3 \log 2\ k \log n/k)$ expected time. Essentially, both procedures create a simplicial partition, which is defined as a set of disconnected triangles whose union includes all of the point data and such that each triangle is consistent with the needed precision. A TIN is formed by completing a set of triangles of this type. A simpler version of the latter algorithm has produced empirical findings that are equivalent to those achieved with other approaches. Finding a simplicial partition necessitates the use of advanced techniques, the implementation of which is difficult in reality (Gorokhovich and Voustianiouk, 2006; Wu et al., 2013).

Silva et al. (1994) have suggested an algorithm that resolves the first problem by employing a greedy strategy. When building an approximation TIN, iteratively cutting triangles from hollow polygons spanning the domain is used: at every step, the algorithm evaluates a bites and polygon from it a triangle with the largest area that is similar with the needed precision. In addition, it is hypothesized that this approach may be able to ensure a bound on the size of the answer; however, demonstrating this is currently a work in progress. This technique works with straightforward data structures and needs only a tiny amount of memory to operate.

When compared to heuristic-based methods, the majority of other algorithms employ iterative methods in which they attempt to select a "good" subset of the input dataset to serve as a collection of vertices for the approximate model. These approaches can be employed for either the first or second problem, depending on the test that is used to determine when the iteration should be stopped. Fine-tuning heuristics begin with a grid of vertices that represents a very small portion of the input data. In this process, new data points are continuously inserted as vertices of the model until all of the essential constraints have been met. In contrast, simplification heuristics act in the opposite way by iteratively removing points from an original model that was constructed over the entire data set. In this scenario, points are selected at every iteration in order to cause the error to increase by the least amount possible. Techniques for extracting morphological features from an initial dataset have also been used by some writers to establish constraints for the approximation TIN (Florinsky, 2002).

Because of their irregular structure, TINs are significantly more versatile than estimate RSGs, which can only be generated by subsampling the grid. From a given TIN, a plethora of approximation models with varying degrees of accuracy and sizes can be constructed. When it comes to contour maps, there are just a few options for varying the resolution by eliminating lines, whereas line simplification allows for a far wider variety of resolutions and data sizes to be obtained.

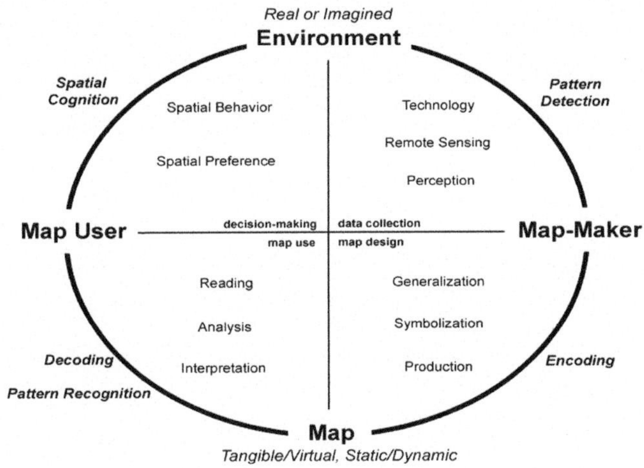

Figure 5.19. The cartographic process.

Source: https://www.e-education.psu.edu/geog160/node/1882.

Methods of refinement. In Ref. Florinsky (2002), Fowler and Little suggest a technique that starts with an RSG and builds an approximation TIN in two steps, beginning with the RSG and ending with the approximate TIN.

Initial TIN construction begins with the selection of and the construction of an initial TIN based on the 3D shape of such points. The model is then developed by iteratively adding the point corresponding to the largest approximation error and updating the 3D mesh, which is a process known as iterative addition. In addition, Franklin devised a comparable technique that does not use point features, and an effective implementation of this algorithm is accessible in the public domain (Figure 5.19) (Florinsky, 2002).

Simplification methods – Lee presented a Delaunay triangulation-based simplification approach. Over the entire dataset, a Delaunay TIN is created initially. The TIN's vertices are then iteratively eliminated, using a criterion akin to Fowler and Little's: at every iteration, the point whose loss produces the minimum increase in error is dropped, and the triangle is updated accordingly.

To develop a multi-resolution model, de Berg and Dobrindt (1998) presented another simplification approach based on Kirk-hierarchical Patrick's triangulation scheme. At each iteration, a maximum number of independent vertices is deleted, and holes are filled using a Delaunay triangulation. De Floriani et al. (1994) studied and compared variants of the above algorithms.

5.4.3. Terrain Analysis

Complex studies of terrain models include (Florinsky, 2002):

- Solution of visibility-related problems and computation of visibility maps;
- Terrain simplification, and feature extraction;
- Computation of crisis, and drainage systems;
- Realistic vision;
- Path planning.

For developing approximated and multiresolution models, terrain generalization difficulties were addressed in Section 4.3. In Section 5.2 on topographic features, related topics will be discussed.

Methods for orthographic presentation and perspective display, along with advanced visualization techniques, are included in the translation of DTMs and associated information (Figure 5.20).

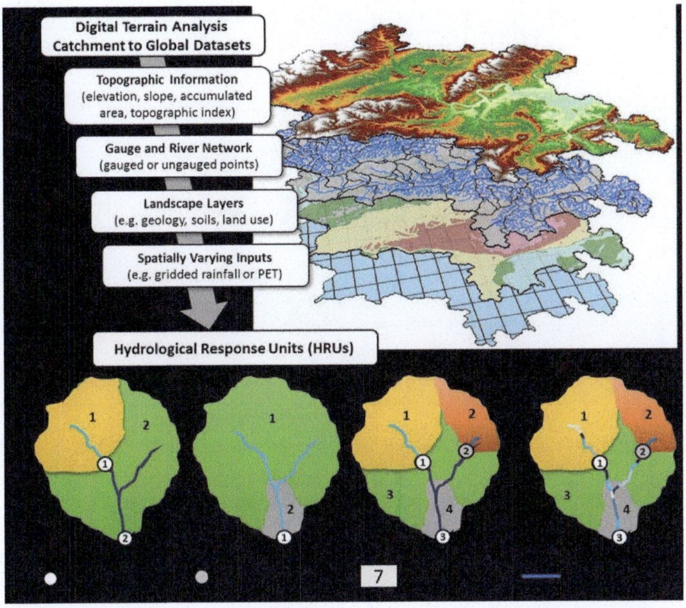

Figure 5.20. Digital terrain analysis.

Note: Simplified examples of applying classification layers to divide a hypothetical watershed into hydrological response units using digital terrain analysis.

5.5. THREE-DIMENSIONAL GIS

Terrains are not actual 3D objects in traditional GISs because most reasoning and modeling takes place in their two-dimensional realm. Adding genuine three-dimensionality to GIS is a more recent study field, and it involves computational geometry directly.

In cadastral and architectural applications, 3D abilities in a GIS system are necessary to represent buildings, while in geosciences and environmental modeling, three-dimensional capabilities are required to represent the structure of either the earth or the atmosphere. Techniques inherited from three-dimensional modeling and CAD systems can be used to handle three-dimensional geographic data.

The shape of man-made substances is typically regular and well-known, allowing them to be readily represented using traditional CAD techniques. Complex structures are known just at a finite set of sampling points when modeling physical processes (e.g., in geology), and interpolation must be employed to recover a complete model of a 3D area. In one dimension, the scenario is comparable to that of landscape modeling. In geology, Jones (2004) gives an outline of data models for tri objects (Figure 5.21).

Figure 5.21. 3D image with digitized GIS features.

Source: https://www.researchgate.net/figure/3D-image-with-Digitized-GIS-features-Figure-13-shows-the-terrain-in-three-dimensional_fig6_224831011.

While two-dimensional domains can simply be sampled uniformly and densely. Because of the high expenses, three-dimensional sampling is frequently sparse. As a result, a typical three-dimensional grid would be a waste of memory. More appropriate are adaptive regular grids, like octrees and their derivatives. The octree is a 3D quadtree generalization. It's centered on a recursive division of a cubic world into nested cubes of various sizes that may be unfurled into an eight-degree tree. An attributes value is assigned to each cubic cell. The need for a regular distribution of data points, as well as the rough calculation of boundaries between zones characterized by diverse attribute values, are disadvantages of this representation: this is not acceptable when data about the surface trying to separate two different attributes' must be maintained in the 3D model.

REFERENCES

1. Agarwal, P. K., & Desikan, P. K., (2000). Approximation algorithms for layered manufacturing. In: SODA (Vol. 1, pp. 528–537).
2. Agarwal, P. K., Van, K. M., & Suri, S., (1998). Label placement by maximum independent set in rectangles. Computational Geometry, 11(3, 4), 209–218.
3. Albert, W. S., & Golledge, R. G., (1999). The use of spatial cognitive abilities in geographical information systems: The map overlay operation. Transactions in GIS, 3(1), 7–21.
4. Allen, G. I., & Tibshirani, R., (2012). Inference with transposable data: Modeling the effects of row and column correlations. Journal of the Royal Statistical Society: Series B (Statistical Methodology), 74(4), 721–743.
5. Alrashed, A. A., Gharibdousti, M. S., Goodarzi, M., De Oliveira, L. R., Safaei, M. R., & Bandarra, F. E. P., (2018). Effects on thermophysical properties of carbon based nanofluids: Experimental data, modeling using regression, ANFIS and ANN. International Journal of Heat and Mass Transfer, 125, 920–932.
6. Amadi, C., Leary, R., Palmeri, A., Hedley, V., Sen, A., Siddiqui, R. Q., & Straub, V., (2022). How can a clinical data modeling tool be used to represent data items of relevance to pediatric clinical trials? Learning from the conect4children (c4c) consortium. Applied Sciences, 12(3), 1604.
7. Arayici, Y., (2007). An approach for real world data modeling with the 3D terrestrial laser scanner for built environment. Automation in Construction, 16(6), 816–829.
8. Arbia, G., Griffith, D., & Haining, R., (1998). Error propagation modeling in raster GIS: Overlay operations. International Journal of Geographical Information Science, 12(2), 145–167.
9. Azab, A. A., (2014). Agnes high, Western Desert, Egypt: A structural study in view of potential data modeling. Egyptian Journal of Petroleum, 23(2), 229–245.
10. Barreira-González, P., Gómez-Delgado, M., & Aguilera-Benavente, F., (2015). From raster to vector cellular automata models: A new approach to simulate urban growth with the help of graph theory. Computers, Environment and Urban Systems, 54, 119–131.
11. Batra, D., & Davis, J. G., (1992). Conceptual data modeling in database design: Similarities and differences between expert and novice designers. International Journal of Man-Machine Studies, 37(1), 83–101.

12. Batra, D., & Marakas, G. M., (1995). Conceptual data modeling in theory and practice. European Journal of Information Systems, 4(3), 185–193.
13. Battersby, S. E., Golledge, R. G., & Marsh, M. J., (2006). Incidental learning of geospatial concepts across grade levels: Map overlay. Journal of Geography, 105(4), 139–146.
14. Berg, D. E., Kohn, M. A., Farley, T. A., & McFarland, L. M., (2000). Multi-state outbreaks of acute gastroenteritis traced to fecal-contaminated oysters harvested in Louisiana. The Journal of Infectious Diseases, 181(Supplement_2), S381–S386.
15. Berg, M. D., Kreveld, M. V., Overmars, M., & Schwarzkopf, O., (1997). Computational geometry. In: Computational Geometry (pp. 1–17). Springer, Berlin, Heidelberg.
16. Bonham-Carter, G. F., & Bonham-Carter, G., (1994). Geographic Information Systems for Geoscientists: Modeling with GIS (No. 13). Elsevier.
17. Bornberg-Bauer, E., & Paton, N. W., (2002). Conceptual data modeling for bioinformatics. Briefings in Bioinformatics, 3(2), 166–180.
18. Brassel, K. E., & Weibel, R., (1988). A review and conceptual framework of automated map generalization. International Journal of Geographical Information System, 2(3), 229–244.
19. Brisaboa, N. R., Bernardo, G. D., Gutiérrez, G., Luaces, M. R., & Paramá, J. R., (2017). Efficiently querying vector and raster data. The Computer Journal, 60(9), 1395–1413.
20. Buttenfield, B. P., & McMaster, R. B., (1991). Map Generalization: Making Rules for Knowledge Representation (p. 244). London: Longman Scientific & Technical.
21. Câmara, G., Souza, R. C. M., Freitas, U. M., & Garrido, J., (1996). SPRING: Integrating remote sensing and GIS by object-oriented data modeling. Computers & Graphics, 20(3), 395–403.
22. Chauve, A., Mallet, C., Bretar, F., Durrieu, S., Pierrot-Deseilligny, M., & Puech, W., (2007). Processing full-waveform lidar data: Modeling raw signals. In: ISPRS Workshop Laser Scanning and SilviLaser (LS SL) (Vol. 36, pp. 102–107).
23. Costa, E., Costa, C., & Santos, M. Y., (2017). Efficient big data modeling and organization for Hadoop hive-based data warehouses. In: European, Mediterranean, and Middle Eastern Conference on Information Systems (pp. 3–16). Springer, Cham.
24. Cromley, R. G., & Hanink, D. M., (1999). Coupling land use allocation models with raster GIS. Journal of Geographical Systems, 1(2), 137–153.

25. Da Paz, A. R., Collischonn, W., Risso, A., & Mendes, C. A. B., (2008). Errors in river lengths derived from raster digital elevation models. Computers & Geosciences, 34(11), 1584–1596.
26. De Berg, M., & Dobrindt, K. T., (1998). On levels of detail in terrains. Graphical Models and Image Processing, 60(1), 1–12.
27. De Floriani, L., Magillo, P., & Puppo, E., (2000). Applications of computational geometry to geographic information systems. Handbook of Computational Geometry, 7, 333–388.
28. De Floriani, L., Marzano, P., & Puppo, E., (1994). Line-of-sight communication on terrain models. International Journal of Geographical Information Systems, 8(4), 329–342.
29. De Silva, A. P., Gunaratne, H. N., Gunnlaugsson, T., Huxley, A. J., McCoy, C. P., Rademacher, J. T., & Rice, T. E., (1997). Signaling recognition events with fluorescent sensors and switches. Chemical Reviews, 97(5), 1515–1566.
30. Dyn, N., Levin, D., & Rippa, S., (1990). Data dependent triangulations for piecewise linear interpolation. IMA Journal of Numerical Analysis, 10(1), 137–154.
31. Egenhofer, M. J., & Richards, J. R., (1993). Exploratory access to geographic data based on the map-overlay metaphor. Journal of Visual Languages & Computing, 4(2), 105–125.
32. Emsley, M. W., Lowe, D. J., Duff, A. R., Harding, A., & Hickson, A., (2002). Data modeling and the application of a neural network approach to the prediction of total construction costs. Construction Management & Economics, 20(6), 465–472.
33. Fang, W. U., Xianyong, G. O. N. G., & Jiawei, D. U., (2017). Overview of the research progress in automated map generalization. Acta Geodaetica et Cartographica Sinica, 46(10), 1645.
34. Fisher, P., & Unwin, D., (2005). Re-Presenting Geographical Information Systems (pp. 1–17). Re-presenting GIS, London: Wiley.
35. Floriani, L. D., Magillo, P., & Puppo, E., (2000). VARIANT: A system for terrain modeling at variable resolution. GeoInformatica, 4(3), 287–315.
36. Florinsky, I. V., (2002). Errors of signal processing in digital terrain modeling. International Journal of Geographical Information Science, 16(5), 475–501.
37. Fonseca, Í. A., & Gaspar, H. M., (2021). Challenges when creating a cohesive digital twinship: A data modeling perspective. Ship Technology Research, 68(2), 70–83.

38. Frank, A. U., (1988). Requirements for a database management system for a GIS. Photogramm. Eng. Remote Sens., 54(11), 1557–1564.
39. Fullea, J., Rodríguez-González, J., Charco, M., Martinec, Z., Negredo, A., & Villaseñor, A., (2015). Perturbing effects of sub-lithospheric mass anomalies in GOCE gravity gradient and other gravity data modeling: Application to the Atlantic-Mediterranean transition zone. International Journal of Applied Earth Observation and Geoinformation, 35, 54–69.
40. Gaeuman, D., Symanzik, J., & Schmidt, J. C., (2005). A map overlay error model based on boundary geometry. Geographical Analysis, 37(3), 350–369.
41. Gold, C. M., (1994). Three approaches to automated topology, and how computational geometry helps. In: Proceedings, Sixth International Symposium on Spatial Data Handling (pp. 145–158).
42. Gorokhovich, Y., & Voustianiouk, A., (2006). Accuracy assessment of the processed SRTM-based elevation data by CGIAR using field data from USA and Thailand and its relation to the terrain characteristics. Remote sensing of Environment, 104(4), 409–415.
43. Guibas, L., & Stolfi, J., (1985). Primitives for the manipulation of general subdivisions and the computation of Voronoi. ACM transactions on Graphics (TOG), 4(2), 74–123.
44. Hadzilacos, T., & Tryfona, N., (1996). Logical data modeling for geographical applications. International Journal of Geographical Information Science, 10(2), 179–203.
45. Held, M., & Williamson, R. B., (2004). Creating electrical distribution boundaries using computational geometry. IEEE Transactions on Power Systems, 19(3), 1342–1347.
46. Herbei, M., Ular, R., & Dragomir, L., (2011). Map Overlay in GIS (No. 56, pp. 70). Transactions on Hydrotechnics, Politehnica University Timisoara.
47. Herrera, G., Fernández, J. A., Tomás, R., Cooksley, G., & Mulas, J., (2009). Advanced interpretation of subsidence in Murcia (SE Spain) using A-DInSAR data-modeling and validation. Natural Hazards and Earth System Sciences, 9(3), 647–661.
48. Holroyd, F., & Bell, S. B., (1992). Raster GIS: Models of raster encoding. Computers & Geosciences, 18(4), 419–426.
49. Huang, Y., Zhu, D., Lv, Q., Porter, A. L., Robinson, D. K., & Wang, X., (2017). Early insights on the emerging sources citation index (ESCI): An overlay map-based bibliometric study. Scientometrics, 111(3), 2041–2057.
50. Imai, K., (2016). Computational geometry and its applications to

geographic information system. IEICE Technical Report; IEICE Tech. Rep., 115(510), 21–31.

51. Jenness, J. S., (2004). Calculating landscape surface area from digital elevation models. Wildlife Society Bulletin, 32(3), 829–839.
52. Jones, C. B., Bundy, G. L., & Ware, M. J., (1995). Map generalization with a triangulated data structure. Cartography and Geographic Information Systems, 22(4), 317–331.
53. Jones, G., Snead, N., & Nelson, P., (2004). Geology and wine 8. Modeling viticultural landscapes: A GIS analysis of the terroir potential in the Umpqua Valley of Oregon. Geoscience Canada, 31(4), 167–178.
54. Kemp, K. K., Goodchild, M. F., & Dodson, R. F., (1992). Teaching GIS in geography. The Professional Geographer, 44(2), 181–191.
55. Kriegel, H. P., Brinkhoff, T., & Schneider, R., (1991). An efficient map overlay algorithm based on spatial access methods and computational geometry. In: Proc. Int. Workshop on Database Management Systems for Geographical Applications (pp. 194–211). Capri, Italy.
56. Kriegel, H. P., Brinkhoff, T., & Schneider, R., (1991). The combination of spatial access methods and computational geometry in geographic database systems. In: Symposium on Spatial Databases (pp. 5–21). Springer, Berlin, Heidelberg.
57. Kuitert, W., (2013). Urban landscape systems understood by geo-history map overlay. Journal of Landscape Architecture, 8(1), 54–63.
58. Li, X., Wang, W., Li, F., & Deng, X., (1999). GIS based map overlay method for comprehensive assessment of road environmental impact. Transportation Research Part D: Transport and Environment, 4(3), 147–158.
59. Li, Z., (2007). Digital map generalization at the age of enlightenment: A review of the first forty years. The Cartographic Journal, 44(1), 80–93.
60. Livingstone, D. J., Manallack, D. T., & Tetko, I. V., (1997). Data modeling with neural networks: Advantages and limitations. Journal of Computer-Aided Molecular Design, 11(2), 135–142.
61. Lomov, S. V., Ivanov, D. S., Verpoest, I., Zako, M., Kurashiki, T., Nakai, H., & Hirosawa, S., (2007). Meso-FE modeling of textile composites: Road map, data flow and algorithms. Composites Science and Technology, 67(9), 1870–1891.
62. Mackaness, W. A., & Beard, K. M., (1993). Use of graph theory to support map generalization. Cartography and Geographic Information Systems, 20(4), 210–221.

63. Maffini, G., (1987). Raster versus vector data encoding and handling: A commentary. Photogrammetric Engineering and Remote Sensing, 53(10), 1397–1398.
64. Magalhães, S. V., Andrade, M. V., Franklin, W. R., & Li, W., (2015). Fast exact parallel map overlay using a two-level uniform grid. In: Proceedings of the 4th International ACM SIGSPATIAL Workshop on Analytics for Big Geospatial Data (pp. 45–54).
65. Mark, D. M., (2003). Geographic information science: Defining the field. Foundations of Geographic Information Science, 1, 3–18.
66. McKeown, D. M., (1987). The role of artificial intelligence in the integration of remotely sensed data with geographic information systems. IEEE Transactions on Geoscience and Remote Sensing, (3), 330–348.
67. Mendonça-Santos, M. D. L., Santos, H. D., Dart, R. O., & Pares, J. G., (2008). Digital mapping of soil classes in Rio de Janeiro State, Brazil: Data, modeling and prediction. In: Digital Soil Mapping with Limited Data (pp. 381–396). Springer, Dordrecht.
68. Miller, H. J., & Wentz, E. A., (2003). Representation and spatial analysis in geographic information systems. Annals of the Association of American Geographers, 93(3), 574–594.
69. Milton, S. K., & Kazmierczak, E., (2004). An ontology of data modeling languages: A study using a common-sense realistic ontology. Journal of Database Management (JDM), 15(2), 19–38.
70. Newcomer, J. A., & Szajgin, J., (1984). Accumulation of thematic map errors in digital overlay analysis. The American Cartographer, 11(1), 58–62.
71. Overmars, M. H., (1996). Designing the computational geometry algorithms library CGAL. In: Workshop on Applied Computational Geometry (pp. 53–58). Springer, Berlin, Heidelberg.
72. Papy, J. M., De Lathauwer, L., & Van, H. S., (2007). A shift invariance-based order-selection technique for exponential data modeling. IEEE Signal Processing Letters, 14(7), 473–476.
73. Peuquet, D. J., (1984). A conceptual framework and comparison of spatial data models. Cartographica: The International Journal for Geographic Information and Geovisualization, 21(4), 66–113.
74. Regnauld, N., (2001). Contextual building typification in automated map generalization. Algorithmica, 30(2), 312–333.
75. Richards, J., & Egenhofer, M., (1995). A comparison of two direct-manipulation GIS user interfaces for map overlay. Geographical Systems, 2(4), 267–290.

76. Saalfeld, H., & Wedde, M., (1974). Refinement of the crystal structure of gibbsite, A1 (OH) 3. Zeitschrift für Kristallographie-Crystalline Materials, 139(1–6), 129–135.

77. Samet, H., (1990). The Design and Analysis of Spatial Data Structures (Vol. 85, p. 87). Reading, MA: Addison-Wesley.

78. Shahzad, H., Farid, H. U., Khan, Z. M., Anjum, M. N., Ahmad, I., Chen, X., & Gulakhmadov, A., (2020). An integrated use of GIS, geostatistical and map overlay techniques for Spatio-temporal variability analysis of groundwater quality and level in the Punjab province of Pakistan, South Asia. Water, 12(12), 3555.

79. Shekhar, S., Ravada, S., Chubb, D., & Turner, G., (1998). Declustering and load-balancing methods for parallelizing geographic information systems. IEEE Transactions on Knowledge and Data Engineering, 10(4), 632–655.

80. Smith, T. R., Menon, S., Star, J. L., & Estes, J. E., (1987). Requirements and principles for the implementation and construction of large-scale geographic information systems. International Journal of Geographical Information System, 1(1), 13–31.

81. Song, B., Yan, B., Triulzi, G., Alstott, J., & Luo, J., (2019). Overlay technology space map for analyzing design knowledge base of a technology domain: The case of hybrid electric vehicles. Research in Engineering Design, 30(3), 405–423.

82. Sousa, V., & Hey, J., (2013). Understanding the origin of species with genome-scale data: Modeling gene flow. Nature Reviews Genetics, 14(6), 404–414.

83. Stoter, J., Burghardt, D., Duchêne, C., Baella, B., Bakker, N., Blok, C., & Schmid, S., (2009). Methodology for evaluating automated map generalization in commercial software. Computers, Environment and Urban Systems, 33(5), 311–324.

84. Taillandier, P., Banos, A., Drogoul, A., Gaudou, B., Marilleau, N., & Truong, Q. C., (2016). Simulating urban growth with raster and vector models: A case study for the city of Can Tho, Vietnam. In: International Conference on Autonomous Agents and Multiagent Systems (pp. 154–171). Springer, Cham.

85. Trantham, G., & Kennelly, P., (2022). Terrain representation using orientation. Cartography and Geographic Information Science, 1–13.

86. Van, K. M., Van, O. R., Bajaj, C., Pascucci, V., & Schikore, D., (1997). Contour trees and small seed sets for isosurface traversal. In: Proceedings of the Thirteenth Annual Symposium on Computational Geometry (pp. 212–220).

87. Van, O. P., (1994). An R-tree based map-overlay algorithm. In: Proc. EGIS (Vol. 94, pp. 318–327).
88. Veregin, H., (1989). Error modeling for the map overlay operation. Accuracy of Spatial Databases, 290.
89. Verstraete, J., (2015). Solving the map overlay problem with a fuzzy approach. In: Uncertainties in Greenhouse Gas Inventories (pp. 141–154). Springer, Cham.
90. Wang, X., Zhou, X., & Lu, S., (2000). Spatiotemporal data modeling and management: A survey. In: Proceedings 36th International Conference on Technology of Object-Oriented Languages and Systems. TOOLS-Asia 2000 (pp. 202–211). IEEE.
91. Ware, J. M., & Jones, C. B., (1998). Conflict reduction in map generalization using iterative improvement. GeoInformatica, 2(4), 383–407.
92. Wassan, J. T., (2015). Discovering big data modeling for the educational world. Procedia-Social and Behavioral Sciences, 176, 642–649.
93. Wickens, C. D., Kroft, P., & Yeh, M., (2000). Database overlay in electronic map design: Testing a computational model. In: Proceedings of the Human Factors and Ergonomics Society Annual Meeting (Vol. 44, No. 21, pp. 3–451). Sage CA: Los Angeles, CA: SAGE Publications.
94. Wise, S., (2000). GIS data modeling lessons from the analysis of DTMs. International Journal of Geographical Information Science, 14(4), 313–318.
95. Worboys, M. F., Hearnshaw, H. M., & Maguire, D. J., (1990). Object-oriented data modeling for spatial databases. International Journal of Geographical Information System, 4(4), 369–383.
96. Wu, J., Li, Q., He, M., & Zhang, F., (2013). A terrain model generation method based on a 2D plan laser scanner for micro-UAV autonomous flight. In: Proceedings of the 2012 International Conference on Information Technology and Software Engineering (pp. 215–225). Springer, Berlin, Heidelberg.
97. Wu, P. Y., & Franklin, W. R., (1990). A logic programming approach to cartographic map overlay. Computational Intelligence, 6(2), 61–70.

CHAPTER

6 Robot Motion Planning

Contents

6.1. INTRODUCTION

One of robotics' final ambitions is to develop intelligent robots; the robots that can be instructed about what to do without even being instructed on how to use it. This entails, along with many other things, that a robot must be capable of planning its own movements (Latombe, 2012).

To plot a move, a robot needs to have a few understandings of the area under that it is operating. For instance, a moveable robot going through a factory have to be aware of its surroundings. A floor plan may offer any of this information—the locations of walls and machinery. The robot would have to depend on its sensors for more information. It should be capable of detecting barriers that are not shown on the floor plan—for example, humans (Laumond, 1998). The robot must navigate to its destination location without interacting with either of the barriers that use the knowledge about the surroundings (Figure 6.1).

Figure 6.1. Robotics motion planning simulation.

Source: https://robodk.com/blog/robot-motion-planning-made-easy/.

This challenge of motion planning must be addressed anytime a robot wishes to operate in physical environment. The preceding explanation presupposed the presence of an autonomous robot in an industrial setting. That type of robot still is fairly uncommon in comparison to the robotic systems presently frequently used in industry (Elbanhawi and Simic, 2014).

A robot arm, also known as an articulated robot, is made up of a series of linkages linked by joints. Typically, one side of the arm—its platform securely attached to the surface, whereas the other side brings a hand or a few other types of tools. The number of connections ranges between three and six, and sometimes even more (Barraquand and Latombe, 1991). Typically, there are two kinds of joints: revolute joints, which enable the links to revolve around the joint, similar to an elbow, and prismatic joints, which allow one of the linkages to slip in and out. Robot arms are typically used to build or handle objects, as well as execute operations such as welding or spraying. To accomplish this, they have to be capable of moving between positions without clashing with the surroundings, the thing on which they are working, or—an intriguing complication—with themselves (Figure 6.2) (Tang et al., 2012).

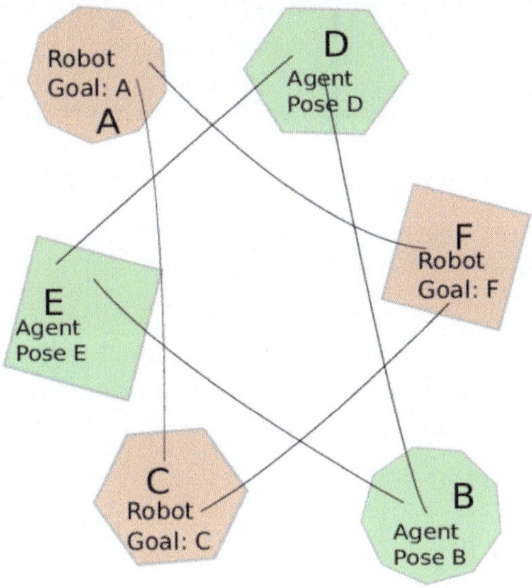

Figure 6.2. Robot motion planning follow-chart.

Source: https://www.researchgate.net/figure/Robot-motion-planning_fig1_319952910.

This chapter discusses some of the fundamental methods and theories consider in motion development. Because the issue of common motion planning is relatively hard, we will end up making a few assumptions (Lengyel et al., 1990).

The most radical simplicity will be to consider a two-dimensional motion planning issue. The surroundings will be a planar geographic area with polygonal obstacles, as will the robot. Additionally, we suppose that now the atmosphere is static—that no one is in the manner of our robot—and that it is defined to the robot (Ichter et al., 2018). The constraint to planar robots is also not as extreme as it looks: for a robot trying to move about on a work floor, a floor plan indicating a position of walls, tables, and machines is frequently enough to design a motion.

The kinds of motions that a robot is capable of executing are determined by its mechanics. Although a few robots are capable of moving in any way, others are restricted in their movements (Tsianos et al., 2007). For example, car-like robots cannot move side to side, parallel parking would've been easier. Additionally, they frequently have a specified least turning radius. Due to the complexity of the geometry governing the movements of vehicle robots, we will limit ourselves to robots capable of moving in any path. Indeed, we will focus exclusively on robots capable of translation (Masehian et al., 2007).

6.2. WORK SPACE AND CONFIGURATION SPACE

Assume 'R' be a robot that moves about in a two-dimensional atmosphere, or work area, that contains a set of obstructions $S = P_1, \ldots, P_t$. 'R' is assumed to be a simple polygon. A translation vector can now be used to specify the robot's placement or setup. 'R' denotes a robot that has been transcribed over a vector (x,y) (x,y) (Burns and Brock, 2005). For example, if the robot is a polygon with vertices (1,1), (1,1), (0,3), (1,1), and (1,1), the vertex shader of R(6,4) are (7,3), (7,5), (6,7), (5,5), and (5,5), respectively (5,3). A robot can be indicated using these expressions by identifying the vertices of R. (0,0).

Another perspective is to think of it as a reference point. If the origin (0,0) is in the inner surface of R, this makes the most sense (0,0). This point is referred to as the robot's reference point by description (Kavraki, 1995). If the robot is in the provided placing, we can clearly state a location reference point of 'R' by merely saying the exact location of the reference point. As a result, R(x,y) clarifies that the robot's reference point is at (x,y). In summary, the point of reference is not on the inside of the robot; this could also be a position just outside of the robot that is connected to it by a hidden stick. This point is the beginning for 'R' by definition (0,0) (Arkani-Hamed et al., 2021).

Assume the robot can rotate along its point of reference to alter its direction. We'll require an extra parameter φ, to define the robot's direction. We'll call the robot R(x,y, φ) since its point of reference is at (x,y) and it rotates anticlockwise over an angle of φ. So, 'R' is the basic specification (0,0,0) (Maciejewski and Fox, 1993).

In generally, a robot's placement is defined by a set of parameters which correlate to the robot's quantity of degrees of freedom (DOF). For planar robots that can just be translated, this quantity is two, while for planar robots which can turn and also translate, it is three (Macho et al., 2009). Of course, the total quantity of parameters required for a robot in three-dimensional space is greater: a translational robot in R^3 has three DOF, while a robot free to translate and rotate in R^3 requires six.

A robot's parameter space is often referred to as its configuration space. 'C' is the abbreviation for it (R). A specific placement R(p) of the robot within work environment correlates to a 'p' in this coordinate space (Figure 6.3) (Curto and Moreno, 1997).

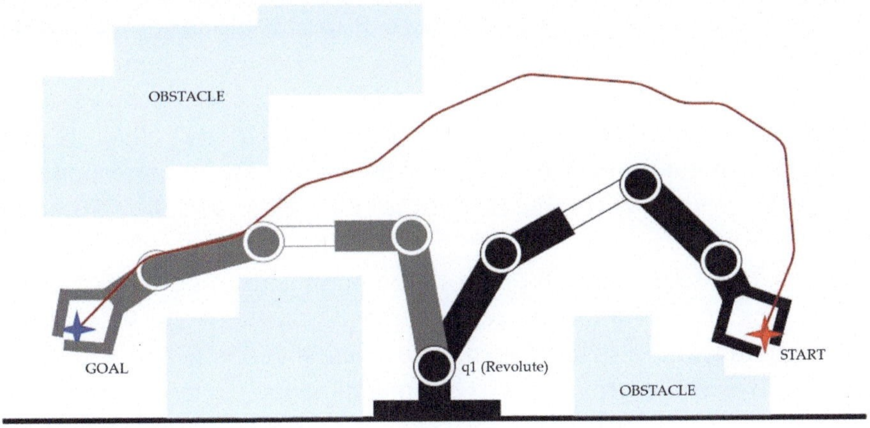

Figure 6.3. Visualization of robot manipulator trajectory planning.

Source: https://control.com/technical-articles/how-does-motion-planning-for-autonomous-robot-manipulation-work/.

The arrangement space for a rotating and translating robot in a plane is three-dimensional. A point (x,y, φ) in this space correlates to the work area placement R(x,y, φ). The configuration space is R^2 [0: 360], which is not the Euclidean three-dimensional space. The configuration space of a rotating robot contains a peculiar topology, which seems to be a cylinder, since rotations across zero and 360° are equal (Lozano-Perez, 1990).

A translating robot's configuration space in the plane is the two-dimensional Euclidean plane, which is equal to the working area. However, it's important to differentiate between the two concepts: the workspace is the area in which the robot keeps moving approximately real world—

and the configuration space is the robot's parameter space. A position in configuration space represents a polygonal robot in the workspace, and each position in configuration space correlates to a specific location of a real robot in the work area (Jaillet et al., 2010). We now contain a mechanism to define a robot position by providing parameters for the factors that determine the location, or, in another saying, by providing a position in configuration space. However, not all locations in configuration space are permissible; points correlating to positions where the robot collides with one of the barriers in 'S' are prohibited (Houben et al., 2014). The region of the configuration space containing these points is known as the banned configuration space, or simply forbidden space. C_{forb} is the symbol for it (R,S). The remainder of the configuration space is known as the free configuration space, or free space, and it is represented by C_{free}. It comprises of the positions corresponding to free placements—placements in which the robot does not collide with any barrier (R,S) (Ivanisevic et al., 2000). A robot's route corresponds to a curve in coordinate space, and vice versa: each position all along path corresponds to a position in configuration space. An arc in free space corresponds to a collision-free route (Wise and Bowyer, 2000). This is seen in Figure 6.4 for a translating planar robot. The work area is depicted on the left, with a collision-free route from the robot's beginning location to its objective position. The configuration space is displayed on the right, with the gray region denoting the restricted area. The open space is the region without shading between the gray areas. The impediments are kept visible in the configuration space for explanation, even if they have no relevance there. Also depicted is the curve that corresponds to the collision-free route (Wu et al., 2009).

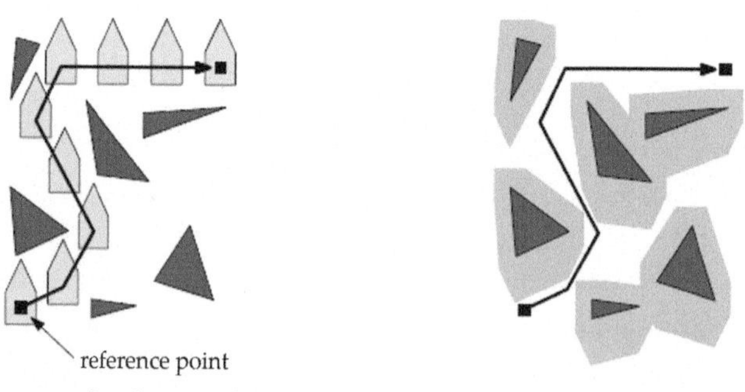

Figure 6.4. A route in the working area and the equivalent curve in the configuration space are shown.

Source: https://studfile.net/preview/5827131/page:14/.

We've seen how to map robot placements to positions in the configuration space, as well as robot tracks to curves in that space. Is it possible to map impediments to configuration space as well? Yes, an obstruction 'P' is transferred to the collection of configuration space points 'p' such that R(p) crosses P. The configuration-space obstacle, or C-obstacle for short, of 'P' is the outcome (Pan and Manocha, 2015).

6.3. POINT ROBOTS

Prior to attempting to arrange the movement of a polygonal robot on a plane, let us examine point robots. Considering the mapping from task space to configuration space discussed above, this is not an outlandish notion. Additionally, it is generally prudent to begin with a straightforward instance (Buckley, 1988). As previously, we will refer to the robot as 'R' and the obstructions as P_1, …, P_t. The obstructions are polygons with discontinuous inners, where 'n' denotes the overall quantity of vertices. The working area and configuration space of a point robot are similar. (That is, assuming we accept the reasonable assumption that the robot's reference point is the robot actually.) Except as then specified, the configuration space is a translation of the workspace (Hwang and Ahuja, 1992).

Rather than attempting to search a way from a given beginning point to a given destination location, we would create a data structure that stores an illustration of the free space. The above data structure could then be used to find the shortest way among any two specified start and destination positions. This strategy is advantageous if the work area of the unimaginably huge obstruction, which would be the area out beyond B, is infinitely large (Tang et al., 2012).

The free space is an unconnected area that might contain holes. Our objective is to calculate an illustration of the free space that enables us to search a way between any two points. This will be accomplished through the use of the trapezoidal map. The trapezoidal map of a collection of un parallel line segments contained within a bounding box is created by creating two vertical additions from each segment node, one upward and one down till a segment (or the reference image) is reached. This approach is used as a subroutine in the proposed method, that calculates a description of the free space (Figure 6.5) (Barfoot and Clark, 2004).

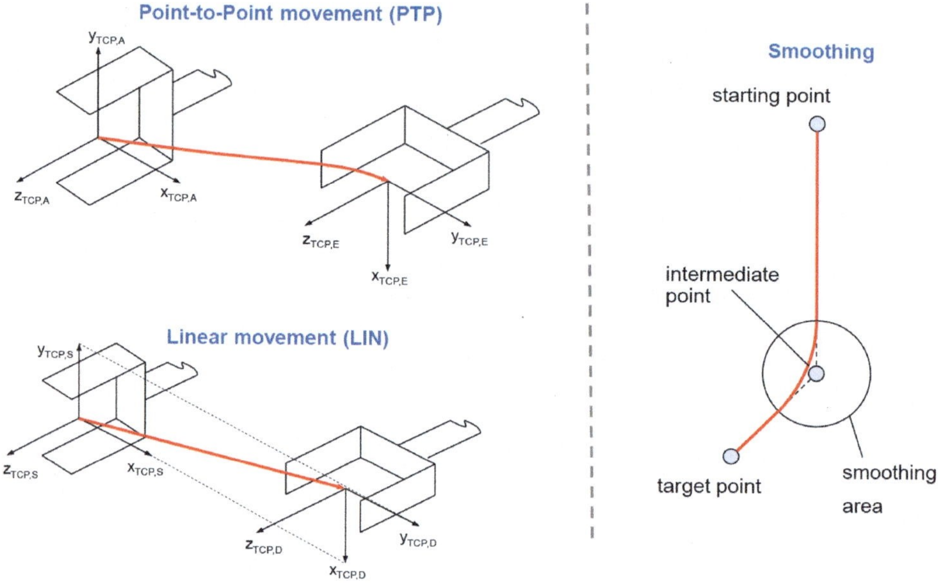

Figure 6.5. Trajectories of the PTP and LIN robot movement orders (left) and smoothing (right).

Source: https://www.researchgate.net/figure/Trajectories-of-robot-movement-commands-PTP-and-LIN-left-and-Smoothing-right_fig1_262391393.

- **COMPUTEFREESPACE(S) Algorithm:**

Input: A collection of disconnected polygons of size S.

Output: C_{free}(R,S) trapezoidal map for a point robot 'R' (Qureshi et al., 2019).

- Define 'E' as the collection of polygonal edges in S;
- Using the method TRAPEZOIDALMAP, calculate the trapezoidal map T(E);
- Subdivision T(E) by removing the trapezoids that are included within one of the polygons.

Figure 6.6 illustrates the method. The figure's part (a) depicts the trapezoidal map of the obstruction edges contained within the bounding box; that's what line 2 of the method computes. Part (b) illustrates the map after the removal of the trapezoids within the obstructions in line 3 (Karami and Hasanzadeh, 2015).

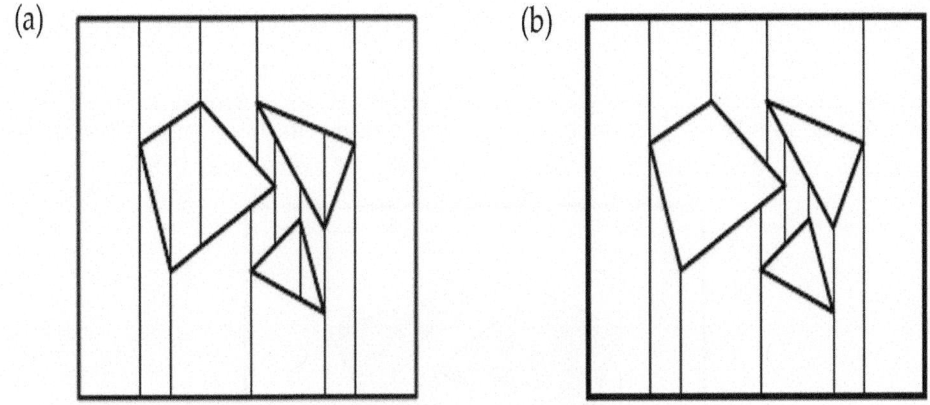

Figure 6.6. Calculation of a trapezoidal map of the open space.

Source: https://studfile.net/preview/5827131/page:14/.

There is one last detail: how can we locate the trapezoids hidden among the obstructions that must be deleted? This is not hard, since after executing TRAPEZOIDALMAP, we understand the edge that surrounds each trapezoid from the top, as well as the obstruction to which that edge relates (Kurniawati et al., 2011). To determine whether to eliminate the trapezoid or not, it is sufficient to determine if the edge confines the obstruction from up or down. The latter test is instantaneous, since the barriers' edges are named in sequence all along the border, so that the obstruction is on a certain, given portion of the edges. However, if the starting and target positions are in separate trapezoids, things get more complicated. In this situation, the route will pass through many trapezoids and may be required to turn in a few of them. We create a road plan across free space to direct the movements over trapezoids (Laumond et al., 1998). The road map is a graph inserted in the plane called G_{road}. It is, more exactly, imbedded in open space. Except for a brief section at the start and end, pathways would always match the road map. Take note as any two adjacent trapezoids have a vertical edge which is an expansion of a segment ending vertically. This brings us to the following road map definition. Each trapezoid has one node in the center, and every vertical extension has one node in the center. An arc occurs among two nodes but only if one node is at the center of a trapezoid and another component is on its perimeter (Ratliff et al., 2009).

Due to the fact that the curves are embedded in the plane as line segments, follow a curve on the road map equates to the robot moving in a straight path

(Lamiraux and Lammond, 2001). This is seen in Figure 6.7. By crossing the doubly-linked edge list of T, the road map G_{road} may be created in O(n) time (C_{free}). We can connect the node in the center of one trapezoid to the node in the core of an adjacent trapezoid through the node on their shared border by using the curves in the road map (Figure 6.8) (Lee and Lee, 1987).

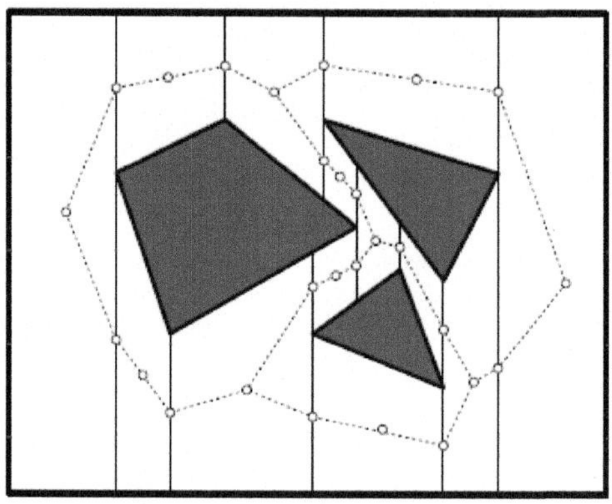

Figure 6.7. A route map is shown.

Source: https://www.amazon.com/Computational-Geometry-Applications-Mark-Berg/dp/3540779736.

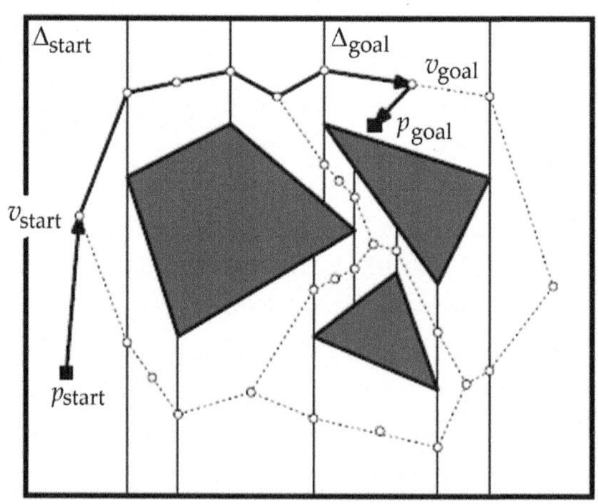

Figure 6.8. A route generated using the road map is shown.

Source: https://www.amazon.com/Computational-Geometry-Applications-Mark-Berg/dp/3540779736.

The following algorithm outlines the process of discovering a route (Masehian and Sedighizadeh, 2007).

- **COMPUTEPATH(T(C_{free}), G_{road}, p_{start}, p_{goal}) Algorithm:**

Input: The open space's trapezoidal map T(C_{free}), a road map G_{road}, a start position p_{start}, and a goal position p_{goal}.

Output: If one exists, a route from p_{start} to p_{goal}. If a route does not exist, it is reported as such (Guo and Parker, 2002).

- Determine the trapezoid start Δ_{start} and the trapezoid goal Δ_{goal} that contain p_{start} and p_{goal}.
- Δ_{start} and Δ_{goal} in the absence of a start or a goal.
- Then indicate that the starting or destination point is in a prohibited area.
- Else Assume that v_{start} is the G_{road} node located in the middle of start.
- Assume that goal is the point of G_{road} located in the middle of goal.
- Using breadth-first search, compute a route in G_{road} from start to destination search.
- In the absence of such a route.
- Then report that no route exists between p_{start} and p_{goal}.
- Else determine the route taken in a straight line from p_{start} to p_{start}, the G_{road} route, and a straight-line path between p_{goal} and p_{goal}.

To use the point position structure described in Chapter 6, it is possible to find the trapezoids comprising the beginning and objective in O(logn) time. Instead, we may just verify all trapezoids in constant time; while the remainder of the procedure requires constant time, it does not significantly raise the time constraint without any symmetry (Canny and Reif, 1987).

The breadth-first find is linear in the graph's size G_{road}. This graph has a node for each trapezoid and a node for each vertical addition. The amount of vertical expansions and trapezoids are both proportional to the total set of vertices in the hurdles. The graph's number of arcs is also linear, since it is planar. As a result, the breadth-first search is O(n) in duration (Latombe, 1999).

The time required to determine a route is determined by the all-out amount of curves that a path may have in G_{road}, which really is O(n) (Belta et al., 2005).

6.4. MINKOWSKI SUMS

We addressed the motion planning issue for a point robot in the previous paragraph; we generated a trapezoidal map of its space available and utilized that map to schedule its moves (Ziegler and Stiller, 2010). If the robot is a polygon, the same method may be employed. One distinction cause working with a polygonal robot extremely complicated: configuration-space impediments are no more just like work-space barriers. As a result, we begin by investigating the open configuration space of a translational polygonal robot (Figure 6.9) (Lengyel et al., 1990).

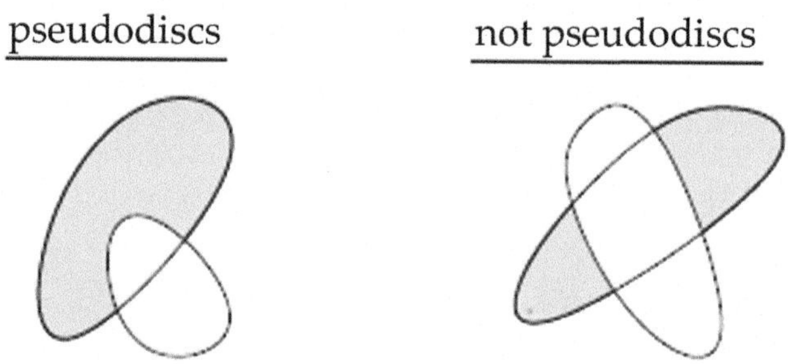

Figure 6.9. The pseudodisc property.

Source: https://www.amazon.com/Computational-Geometry-Applications-Mark-Berg/dp/3540779736.

6.5. TRANSLATIONAL MOTION PLANNING

It's time to come back to the topic of planar motion planning. Remember that our robot 'R' could only move and as the barriers are polygons that really are discontinuous. We already established that the C-obstacle related to an obstruction Pi is the Minkowski sum $P_i(R)$ in the preceding part. Minkowski sums of convex polygons are also pseudo discs, as we've shown (Varadhan and Manocha, 2004). This is used to verify our first significant finding on the motion planning issue, namely that the difficulty of a translational planar robot's free space is linear.

It is still necessary to develop an algorithm for constructing the space available. We'll calculate the prohibited space C_{forb} instead of the free space C_{free}; the free space is merely its counterpart (Lien, 2007).

Let P_1, ..., P_n indicate the triangles formed when the hurdles are triangulated. We want to do some math. We showed how to calculate the individual Minkowski sums CP_i in the previous section. We utilize a divide-and-conquer method to calculate their union (Salzman et al., 2013).

Input: FORBIDDENSPACE(CP_1, …, CP_n) algorithm. A grouping of C-obstacles.

Output: $C_{forb} = \bigcup_{i=1}^{n} CP_i$ is a banned place.

if 'n' equals 1

then go back to CP_1

else C^1_{forb} ←FORBIDDENSPACE $(\mathcal{P}_1, \ldots, \mathcal{P}_{\lceil n/2 \rceil})$4. C^2_{forb} ←FORBIDDENSPACE$(\mathcal{P}_{\lceil n/2 \rceil+1}, \ldots, \mathcal{P}_n)$

$C_{forb} = C_{1forb} \cup C_{2forb}$ is the answer.

C_{forb} should be returned

The subroutine that computes the relationship between two planar areas, which we require to complete the merging step, lies at the core of our approach (line 5). If we describe such areas using edge lists that are doubly linked (Naderan-Tahan and Manzuri-Shalmani, 2009).

6.6. MOTION PLANNING WITH ROTATIONS

Previously, the robot was really only permitted to move. Whenever a robot is circular, its range of motion is unrestricted. On either side, whenever it's long and thin, translation motion alone is frequently insufficient: it may need to alter (Figure 6.10) (Mei et al., 2004).

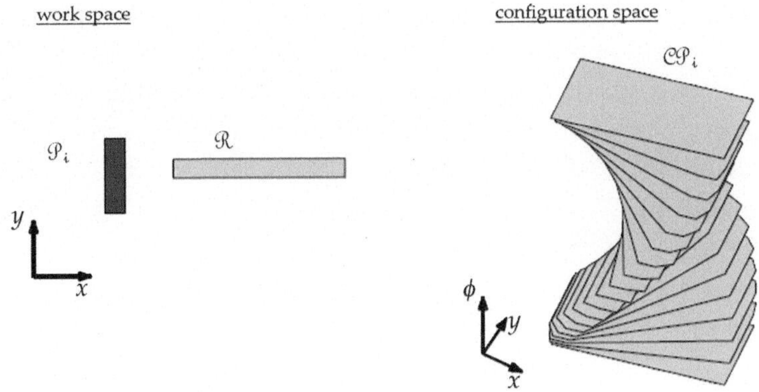

Figure 6.10. A translational and rotational robot's C-obstacle.

Source: https://link.springer.com/book/10.1007/978-3-540-77974-2.

Its direction in order to navigate a narrow passageway or around with an edge. This section describes a method for planning movement for robots which can spin in addition to translating (Lengyel et al., 1990).

Assume that 'R' is a convex polygonal robot capable of translation and rotation within a planar workspace containing a group P_1, ..., P_t of discontinuous polygonal hurdles. Three DOF are available to the robot R: two transitionary and one rotational (Overmars and Svestka, 1994). As a result, we can determine the position of 'R' using three variables: the x- and y-coordinates of a location point for R, as well as an angle φ specifying its alignment.

The configuration space obtained is the three-dimensional space $R^2 \times [0:360)$, with a topology consisting of the points (x,y,0) and (x,y,360). Consider the following definition of CP_i, the C-obstacle of a barrier P_i (Reif and Sharir, 1994):

$CP_{i:} \{(x,y,\phi) \in \mathbb{R}^2 \times [0:360) \ : \ \mathcal{R}(x,y,\phi) \cap \mathcal{P}_i \neq \emptyset\}.$.

How do these C-obstacles appear? This is a tough subject to explicitly answer, but we can gain a sense of it by examining cross-sections with planes of φ constants. Because the rotation angle is constant in just such a plane, we are faced with a strictly translation issue (Overmars, 1992). As a result, we understand the cross-shape: section's it is a Minkowski sum. Specifically, the cross-section of CP_i with the plane h: = $\varphi = \varphi_0$ equals $P_i \oplus R(0,0,\varphi_0)$. (More accurately, it is a zero-height φ_0 replica of the Minkowski sum) (Belta et al., 2005). Now consider sweeping a horizontal line upward across configuration space, from φ = 0 to φ = 360. The cross-section of the plane with CP_i is a Minkowski summation at any point at the time of the sweep. The form of the Minkowski sum varies continuously: at = 0, it is $P_i \oplus R(0,0,\varphi_0)$, whereas at $\varphi = \varphi_0 + \varepsilon$, it is $P_i \oplus R(0,0,\varphi_0 + \varepsilon)$. This results in CPi resembling a twisted pillar, as seen in Figure 6.10. Excluding the top and bottom facets, the sides and aspects of this twisting pillar are bent (Mukherjee et al., 2002).

Thus, we have a general idea of what C-obstacles are as. The open space is the sum of these C-obstacles. Because of the obnoxious form of the C-obstacles, the free space is more difficult: its border is no more polygonal, but arced. Additionally, the open space's combinatorial difficulty may be quadratic for convex robots and even cubic for non-convex robots (Geraerts and Overmars, 2007). Nonetheless, we can tackle the motion planning issue utilizing the similar strategy as previously: decompose the empty area into basic cells and create a road map to direct the movements among adjoining cells (Herman, 1986). Specified the robot's start and destination positions, we then find the following route. We convert these locations to configuration space points, locate the cells containing the points, and build a path made up of three components: a path from of the original location to the road map node in the center of the initial cell, a way along the road map to the endpoint in the center of the goal cell, and a path within the goal cell to the final location. It keeps to translate the configuration space route to a workspace motion (LaValle and Hutchinson, 1998).

Due to the intricate structure of the C-obstacles, it is challenging to calculate an appropriate cell breakdown, much more so in practice. As a result, we will discuss a new, more straightforward way. As we will see, this strategy is not without problems (Canny and Reif, 1987). Our solution is based on the same finding that we made while studying the form of the C-obstacles, which is that the motion planning issue simplifies to a simply translational issue when the configuration space is viewed horizontally. This kind of cross-section will be referred to as a slice. Calculate a limited number of slices. A robot's route now contains two kinds of motion: movements inside a slice—which are entirely translational—and motions between slices—which are purely rotational (Lozano-Pérez and Donnell, 1991).

Allow me to formalize this. Assume 'z' signify the quantity of slices that we are going to take. For any integer 'i' between 0 i z−1, let $\varphi_i = i\times(360/z)$. For each 'I' we calculate a slice of the free space. Because we are working with a strictly translational issue for the robot R(0,0,i) inside the slice, we can calculate it using the techniques described in the preceding section (Gilbert and Johnson, 1985). This produces the trapezoidal map T_i of the slice's free space. We generate a route map G_i for each T_i. These road maps serve as a guide for the movements inside a slice (Kambhampati and Davis, 1986).

The only thing left is to join the slices together. To be more specific, we link each set of roadmaps G_i, G_{i+1} to produce a G_{road} map of whole configuration space. This is accomplished in the following manner. We use the algorithm computes the overlay of every set of consecutive slices' trapezoidal maps (Schwartz and Sharir, 1988). (Technically accurate, we calculate the layer of T_i and T_{i+1} projections onto the plane h: $\varphi = 0$). This tells us about all the pairs Δ_1,Δ_2 that have $\Delta_1 \in T_i$ and $\Delta_2 \in T_i+_1$ and where Δ_1 overlaps Δ_2. Let's say (x,y,0) is a point in $\Delta_1 \cap \Delta_2$. Then, at (x,y,φ_i) and (x,y,φ_{i+1}), we insert an extra node to G_{road}, which we connect with an arc. A rotation from φ_i to φ_{i+1}, or back, is equivalent to having to move through one piece to the next along this arc (Hwang and Ahuja, 1992). Moreover, the node at (x,y, φ_i) is linked to the center of Δ_1, and the node at (x,y, φ_{i+1}) is linked to the center of Δ_2. Because such links are kept within such a slice, they only correlate to translational motions. G_{z1} and G_{z2} are linked in the very same manner, G_0. Mention that pathways in the graph G_{road} coincide to robot pathways which are made up of solely translationary motion (whenever we shifted across an arc linking endpoints in the similar slice) and primarily rotational movement (whenever we shifted along a straight-line connecting node in the similar slice) (whenever we shift with an arc linking nodes in various slices) (Barraquand and Latombe, 1991).

We could use this road map to arrange a motion for 'R' from every start location R(xs_{tart}, y_{start}, φ_{start}) to somewhat goal location R(x_{goal}, y_{goal}, φ_{goal}) once we've built it. To do so, we round the orientations start and goal to the nearest orientation φ_i for which we generated a slice to find the slices closest to the φ_{start} and φ_{goal} location. Within those slices, we find the start and goal trapezoids, which include the start and goal positions (Sedighi et al., 2004). We note that we can't calculate a route if one of such trapezoids doesn't exist since the start or goal point is in the restricted region inside the slice. Alternatively, make v_{start} and goal the road map vertices that have been put in the middle. Utilizing breadth-first searching, we aim to locate a route from start to objective in G_{road}. We observe that we are unable to calculate a motion when there's no route in the graph (Konar et al., 2013). Otherwise, we report a five-part motion: a purely rotational movement from of the initial location to an adjacent slice, a solely translational motion in that slice to the node beginning, a movement that relates to the path from start to goal in G_{road}, a purely translational motion from goal to the final location in that slice (which would be the slice closest to the goal location), and eventually a solely rotary movement to the real goal position (Kwon and Yi, 2012).

This approach is a generalization of the one we used to translate movements, but it has a significant flaw: that's not always right. It may occasionally indicate that a route does not exists when this is not the case. For example, the start location might be in free space, but not inside the closest slice. We state that there really is no route in this situation, which does not have to be accurate. Worse, the pathways we report do not have to be devoid of collisions (Van Henten et al., 2003). The translational movements inside a slice are OK since we addressed the problem perfectly within a slice, but the rotational motions between slices may pose issues: the positions between the two slices are collision-free, but the robot might crash with an object halfway through. Whenever we enhance the quantity of slices, both issues become less frequent, but we can never be confident that the output is right

(Barraquand et al., 1992). This is particularly inconvenient in light of the second issue: we don't want our potentially very costly robot to collide (Figure 6.11).

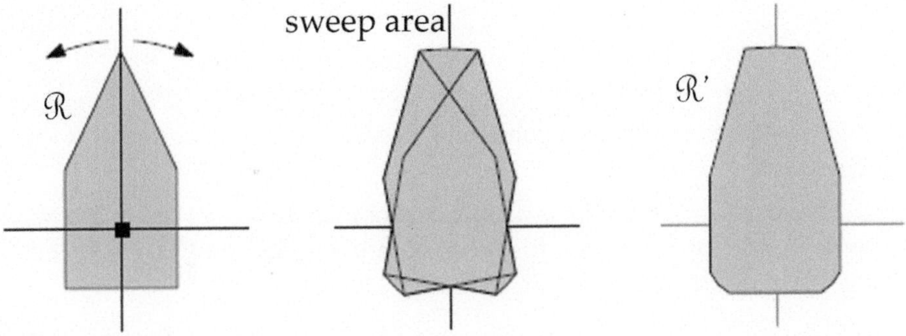

Figure 6.11. Robot enlargement.

Source: https://link.springer.com/book/10.1007/978-3-540-77974-2.

As a result, we employ the following techniques. We extend the robot and apply the procedure outlined above to the bigger robot R. This is performed in a way that, although 'R' may crash with other robots throughout rotations, the initial 'R' cannot. The robot is expanded in the following way to accomplish this. Rotate 'R' in a clockwise and anticlockwise direction across a (180/z)◦ angle (Warren, 1989). During this revolution, 'R' sweeps a part of the plane with notes and comments. Figure 6.11 shows a convex polygon which encompasses the sweep area for the expanded robot R. Instead of R, we calculate the trapezoidal maps and the road map for R. Even though 'R' may hit with an obstruction at the time of a purely rotating motion among two neighboring slices, it is easy to show that 'R' cannot. We have included another approach to mistakenly determine there is no route by extending the robot. Whenever the amount of slices grows, this becomes less probable. As a result, given a high amount of slices, the approach is likely to perform quite well in practice.

REFERENCES

1. Arkani-Hamed, N., Lam, T., & Spradlin, M., (2021). Positive configuration space. Communications in Mathematical Physics, 384(2), 909–954.
2. Barfoot, T. D., & Clark, C. M., (2004). Motion planning for formations of mobile robots. Robotics and Autonomous Systems, 46(2), 65–78.
3. Barraquand, J., & Latombe, J. C., (1991). Robot motion planning: A distributed representation approach. The International Journal of Robotics Research, 1, 10(6), 620–650.
4. Barraquand, J., Langlois, B., & Latombe, J. C., (1992). Numerical potential field techniques for robot path planning. IEEE Transactions on Systems, Man, and Cybernetics, 22(2), 224–241.
5. Belta, C., Isler, V., & Pappas, G. J., (2005). Discrete abstractions for robot motion planning and control in polygonal environments. IEEE Transactions on Robotics, 5, 21(5), 860–875.
6. Buckley, S. J., (1988). Fast Motion Planning for Multiple Moving Robots (Vol. 1, pp. 322–326). IBM Thomas J. Watson Research Division.
7. Burns, B., & Brock, O., (2005). Toward optimal configuration space sampling. In: Robotics: Science and Systems (Vol. 2, pp. 105–112).
8. Canny, J., & Reif, J., (1987). New lower bound techniques for robot motion planning problems. In: 28th Annual Symposium on Foundations of Computer Science (sfcs 1987) (Vol. 2, pp. 49–60). IEEE.
9. Curto, B., & Moreno, V., (1997). Mathematical formalism for the fast evaluation of the configuration space. In: Proceedings 1997 IEEE International Symposium on Computational Intelligence in Robotics and Automation CIRA'97;'Towards New Computational Principles for Robotics and Automation' (Vol. 2, No. 4, pp. 194–199). IEEE.
10. Elbanhawi, M., & Simic, M., (2014). Sampling-based robot motion planning: A review. IEEE Access, 2, 56–77.
11. Geraerts, R., & Overmars, M. H., (2007). Creating high-quality paths for motion planning. The International Journal of Robotics Research, 26(8), 845–863.
12. Gilbert, E., & Johnson, D., (1985). Distance functions and their application to robot path planning in the presence of obstacles. IEEE Journal on Robotics and Automation, 1(1), 21–30.
13. Guo, Y., & Parker, L. E., (2002). A distributed and optimal motion planning approach for multiple mobile robots. In: Proceedings 2002 IEEE International Conference on Robotics and Automation (Cat. No. 02CH37292) (Vol. 3, pp. 2612–2619). IEEE.

14. Herman, M., (1986). Fast, three-dimensional, collision-free motion planning. In: ICRA (Vol. 1, pp. 1056–1063).
15. Houben, S., Tell, P., & Bardram, J. E., (2014). Activity space: Managing device ecologies in an activity-centric configuration space. In: Proceedings of the Ninth ACM International Conference on Interactive Tabletops and Surfaces (Vol. 1, pp. 119–128).
16. Hwang, Y. K., & Ahuja, N., (1992). A potential field approach to path planning. IEEE Transactions on Robotics and Automation, 8(1), 23–32.
17. Ichter, B., Harrison, J., & Pavone, M., (2018). Learning sampling distributions for robot motion planning. In: 2018 IEEE International Conference on Robotics and Automation (ICRA) (Vol. 1, pp. 7087–7094). IEEE.
18. Ivanisevic, I., & Lumelsky, V. J., (2000). Configuration space as a means for augmenting human performance in teleoperation tasks. IEEE Transactions on Systems, Man, and Cybernetics, Part B (Cybernetics), 30(3), 471–484.
19. Jaillet, L., Cortés, J., & Siméon, T., (2010). Sampling-based path planning on configuration-space costmaps. IEEE Transactions on Robotics, 26(4), 635–646.
20. Kambhampati, S., & Davis, L., (1986). Multiresolution path planning for mobile robots. IEEE Journal on Robotics and Automation, 2(3), 135–145.
21. Karami, A. H., & Hasanzadeh, M., (2015). An adaptive genetic algorithm for robot motion planning in 2D complex environments. Computers & Electrical Engineering, 43(1), 317–329.
22. Kavraki, L. E., (1995). Computation of configuration-space obstacles using the fast Fourier transform. IEEE Transactions on Robotics and Automation, 11(3), 408–413.
23. Konar, A., Chakraborty, I. G., Singh, S. J., Jain, L. C., & Nagar, A. K., (2013). A deterministic improved Q-learning for path planning of a mobile robot. IEEE Transactions on Systems, Man, and Cybernetics: Systems, 43(5), 1141–1153.
24. Kurniawati, H., Du, Y., Hsu, D., & Lee, W. S., (2011). Motion planning under uncertainty for robotic tasks with long time horizons. The International Journal of Robotics Research, 30(3), 308–323.
25. Kwon, Y. S., & Yi, B. J., (2012). Design and motion planning of a two-module collaborative indoor pipeline inspection robot. IEEE Transactions on Robotics, 28(3), 681–696.

26. Lamiraux, F., & Lammond, J. P., (2001). Smooth motion planning for car-like vehicles. IEEE Transactions on Robotics and Automation, 17(4), 498–501.
27. Latombe, J. C., (1999). Motion planning: A journey of robots, molecules, digital actors, and other artifacts. The International Journal of Robotics Research, 18(11), 1119–1128.
28. Latombe, J. C., (2012). Robot Motion Planning (Vol. 124, pp. 3–7). Springer Science & Business Media.
29. Laumond, J. P., (1998). Robot Motion Planning and Control (Vol. 229, pp. 1–5). Berlin: Springer.
30. Laumond, J. P., Sekhavat, S., & Lamiraux, F., (1998). Guidelines in nonholonomic motion planning for mobile robots. In: Robot Motion Planning and Control (Vol. 1, pp. 1–53). Springer, Berlin, Heidelberg.
31. LaValle, S. M., & Hutchinson, S. A., (1998). Optimal motion planning for multiple robots having independent goals. IEEE Transactions on Robotics and Automation, 14(6), 912–925.
32. Lee, B. H., & Lee, C. G., (1987). Collision-free motion planning of two robots. IEEE Transactions on Systems, Man, and Cybernetics, 17(1), 21–32.
33. Lengyel, J., Reichert, M., Donald, B. R., & Greenberg, D. P., (1990). Real-time robot motion planning using rasterizing computer graphics hardware. ACM SIGGRAPH Computer Graphics, 24(4), 327–335.
34. Lien, J. M., (2007). Point-based Minkowski sum boundary. In: 15th Pacific Conference on Computer Graphics and Applications (PG'07) (Vol. 2, pp. 261–270). IEEE.
35. Lozano-Pérez, T., & O'Donnell, P. A., (1991). Parallel robot motion planning. In: ICRA (Vol. 1, pp. 1000–1007).
36. Lozano-Perez, T., (1990). Spatial planning: A configuration space approach. In: Autonomous Robot Vehicles (Vol. 1, pp. 259–271). Springer, New York, NY.
37. Macho, E., Altuzarra, O., Amezua, E., & Hernandez, A., (2009). Obtaining configuration space and singularity maps for parallel manipulators. Mechanism and Machine Theory, 44(11), 2110–2125.
38. Maciejewski, A. A., & Fox, J. J., (1993). Path planning and the topology of configuration space. IEEE Transactions on Robotics and Automation, 9(4), 444–456.
39. Masehian, E., & Sedighizadeh, D., (2007). Classic and heuristic approaches in robot motion planning-a chronological review. World Academy of Science, Engineering and Technology, 23(5), 101–106.

40. Mei, Y., Lu, Y. H., Hu, Y. C., & Lee, C. G., (2004). Energy-efficient motion planning for mobile robots. In: IEEE International Conference on Robotics and Automation, 2004; Proceedings. ICRA'04 (Vol. 5, pp. 4344–4349). IEEE.
41. Mukherjee, R., Minor, M. A., & Pukrushpan, J. T., (2002). Motion planning for a spherical mobile robot: Revisiting the classical ball-plate problem. J. Dyn. Sys., Meas., Control, 124(4), 502–511.
42. Naderan-Tahan, M., & Manzuri-Shalmani, M. T., (2009). Efficient and safe path planning for a mobile robot using genetic algorithm. In: 2009 IEEE Congress on Evolutionary Computation (Vol. 1, pp. 2091–2097). IEEE.
43. Overmars, M. H., & Svestka, P., (1994). A Probabilistic Learning Approach to Motion Planning (Vol. 1, pp. 19–37). Department of Computer Science, Utrecht University.
44. Overmars, M. H., (1992). A Random Approach to Motion Planning (Vol. 3, No. 2, pp. 1–3). The Netherlands: Department of Computer Science, Utrecht University.
45. Pan, J., & Manocha, D., (2015). Efficient configuration space construction and optimization for motion planning. Engineering, 1(1), 046–057.
46. Qureshi, A. H., Simeonov, A., Bency, M. J., & Yip, M. C., (2019). Motion planning networks. In: 2019 International Conference on Robotics and Automation (ICRA) (3rd edn., pp. 2118–2124). IEEE.
47. Ratliff, N., Zucker, M., Bagnell, J. A., & Srinivasa, S., (2009). CHOMP: Gradient optimization techniques for efficient motion planning. In: 2009 IEEE International Conference on Robotics and Automation (Vol. 1, pp. 489–494). IEEE.
48. Reif, J., & Sharir, M., (1994). Motion planning in the presence of moving obstacles. Journal of the ACM (JACM), 41(4), 764–790.
49. Salzman, O., Hemmer, M., Raveh, B., & Halperin, D., (2013). Motion planning via manifold samples. Algorithmica, 67(4), 547–565.
50. Schwartz, J. T., & Sharir, M., (1988). A survey of motion planning and related geometric algorithms. Artificial Intelligence, 37(1–3), 157–169.
51. Sedighi, K. H., Ashenayi, K., Manikas, T. W., Wainwright, R. L., & Tai, H. M., (2004). Autonomous local path planning for a mobile robot using a genetic algorithm. In: Proceedings of the 2004 Congress on Evolutionary Computation (IEEE Cat. No. 04TH8753) (Vol. 2, pp. 1338–1345). IEEE.
52. Tang, S. H., Khaksar, W., Ismail, N., & Ariffin, M., (2012). A review on robot motion planning approaches. Pertanika Journal of Science and Technology, 20(1), 15–29.

53. Tsianos, K. I., Sucan, I. A., & Kavraki, L. E., (2007). Sampling-based robot motion planning: Towards realistic applications. Computer Science Review, 1(1), 2–11.

54. Van, H. E. J., Hemming, J., Van, T. B. A. J., Kornet, J. G., & Bontsema, J., (2003). Collision-free motion planning for a cucumber picking robot. Biosystems Engineering, 86(2), 135–144.

55. Varadhan, G., & Manocha, D., (2004). Accurate Minkowski sum approximation of polyhedral models. In: 12th Pacific Conference on Computer Graphics and Applications, 2004; PG 2004; Proceedings (Vol. 3, pp. 392–401). IEEE.

56. Warren, C. W., (1989). Global path planning using artificial potential fields. In: 1989 IEEE International Conference on Robotics and Automation (Vol. 1, pp. 316, 317). IEEE Computer Society.

57. Wise, K. D., & Bowyer, A., (2000). A survey of global configuration-space mapping techniques for a single robot in a static environment. The International Journal of Robotics Research, 19(8), 762–779.

58. Wu, X. J., Tang, J., Li, Q., & Heng, K. H., (2009). Development of a configuration space motion planner for robot in dynamic environment. Robotics and Computer-Integrated Manufacturing, 25(1), 13–31.

59. Yang, S. X., & Meng, M., (2000). An efficient neural network approach to dynamic robot motion planning. Neural Networks, 13(2), 143–148.

60. Ziegler, J., & Stiller, C., (2010). Fast collision checking for intelligent vehicle motion planning. In: 2010 IEEE Intelligent Vehicles Symposium (Vol. 1, pp. 518–522). IEEE.

CHAPTER

7 Application of Computational Geometry in Computer Graphics

Contents

7.1. INTRODUCTION

Computational geometry and computer graphics all explore geometric processes as they pertain to computers. Computational geometry gives a theoretical basis encompassing the research of algorithms and data structures for executing geometric calculations. Computer graphics addresses the realistic design of algorithms, software, and hardware essential to produce graphics (like to show geometry) on the computer screen. On the intersection lay several general difficulties each of that discipline must overcome to attain its aim. Methods in the field are frequently identical, often distinct. And, every field has established a group of paradigms on where to work (Figures 7.1 and 7.2) (Yao, 1990; Sproull, 1991).

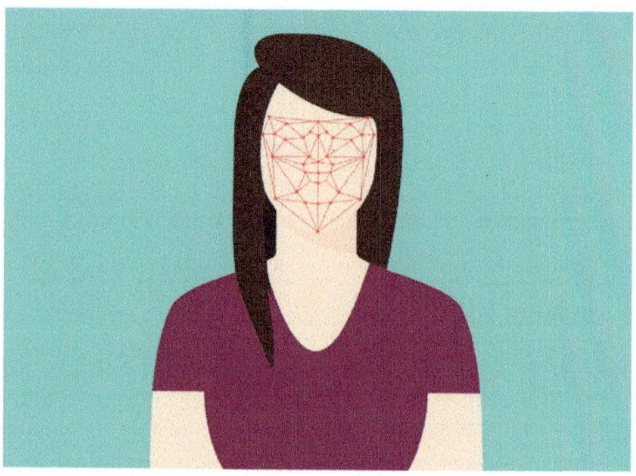

Figure 7.1. Computational geometry is used in machine learning.

Source: https://medium.datadriveninvestor.com/computational-geometry-for-computer-vision-f140fab91c76?gi=f52da88421b5.

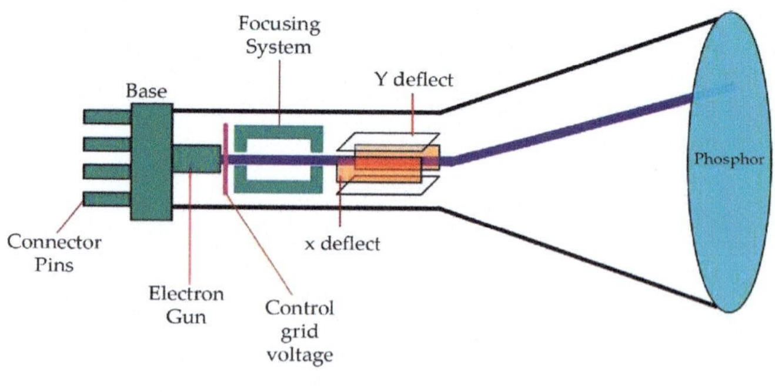

Figure 7.2. Computer graphics basics.

Source: https://www.tutorialspoint.com/computer_graphics/computer_graphics_basics.htm.

7.2. REPRESENTING SUBDIVISIONS OF PLANE AND SPACE

The partition is the most important structure in computational geometry issues. This is the structure being used to hold and describe the interconnections of groups of geometric objects (for example, points, lines, polygons, etc.). These structures are used to solve a variety of fundamental issues. The convex hull, Voronoi diagram, and Delaunay triangulation are by far the most essential of all these (to be explained below).

This topic examines the evolution of subdivision-handling strategies. We'll start with a brief overview of the problem's origins. The mathematical strategies that were utilized to represent the situation are then presented (Samet, 1990). Following that, the data structures for answering the problem are developed. We follow the development of such data structures through the creation of implementable algorithms. The section concludes with a discussion of application areas that use these strategies (Shamos and Hoey, 1976, 1990).

7.2.1. Roots of the Problem

Graham (1972) presented a method in 1971 for finding the convex hull of a group of points in the plane by categorizing and screening the points (Dobkin and Lipton, 1976). His innovation was motivated by the requirement for a Bell Labs statistics department to be capable to combine data samples effectively (Graham, 1972). Dobkin and Lipton (1974) introduced the first methods for exploring inside spatial subdivisions in 1974. Their approach was motivated by an open topic posed in Knuth (1973): how to efficiently pre-process a set of points in order to discover nearest neighbors. Shamos and Hoey introduced efficient methods for determining the nearest couple from a collection of locations in 1975 (Shamos, 1975; Shamos and Hoey, 1975). This research was incorporated into Shamos' dissertation, that established the area of computational geometry. Though it is not recognized at the moment, the challenges of discovering convex hulls, creating, and researching spatial subdivisions, and identifying closest neighbors are all connected. They comprise a lengthy mathematical background and have discovered use in computer graphics in the past years (Samet, 1990).

The Graham algorithm was developed in response to a real-world challenge. At first, spatial searching and nearest point algorithms were merely theoretical. Though prompted by real-world challenges, the search algorithms needed vast quantities of preparation time to construct search structures demanding absurd quantities of memory space. The techniques for computing the nearest points were developed as a reaction to the discoveries of the Voronoi diagram (1908). Techniques for efficiently computing the diagram in a logarithmic logic were presented, however such algorithms were not easily implementable (Peterson, 1984; Salesin et al., 1989).

Scientists in computational geometry have begun researching extensions to the techniques mentioned previously, as well as acceptable data structures for their execution. These developments included strategies that were both efficient and implementable, and also solutions to similar issues in high dimensional or with particular data sets. While this was occurring, Baumgart invented the winged edge data structure (Baumgart, 1976) to aid with his computer vision studies. This structure might be advantageous as a jumping-off point for the implementation techniques discussed below (Paterson and Yao, 1990).

7.2.2. Setting for the Problem

After providing a quick overview of the early the past, I'll introduce a nomenclature which would be utilized throughout this segment. Subsets of Ed, d-dimensional Euclidean space (usually for d=2,3) are considered.

Points, edges, polygons, and polyhedra make up our geometries. The three-dimensional equivalent of a polygon is a polyhedron. Vertices, edges, and faces make up a polyhedron. The faces are made out of polygons. The amount of edges whereby a vertex (resp. face) is close is defined as its degree (according to the amount of edges that compose it). It's worth noting that a polyhedron may be "unfolded" in the plane and extended out. The structure that emerges from this unfold is known as a planar subdivision (Lienhardt, 1989; Max et al., 1990).

We describe the convex hull of the points of 'S' as the shortest convex body encompassing all the points of 'S,' if 'S' is a set of points in Ed. This object is a polygon in the plane. It's a closed polyhedron in three dimensions (known as a polytope). There will be no more as compared to the |S| vertices in it (Haussler and Welzl, 1987; Max et al., 1990).

A spatial subdivision is the partition of Ed into d-dimensional polyhedra in part or whole. Only border crossings are allowed in these polyhedra. Our subdivisions will be further restricted to exclusively convex regions. With the inclusion of new vertices, any subdivision may be turned into one consisting solely of convex areas (see Palios, 1992). As a result, a planar subdivision is the partition of a plane into polygonal parts. Figure 7.3 depicts one example.

It's worth noting that now the vertices and polygons don't have to be of the same degree. A triangulation occurs when all of the polygons in a planar subdivision are triangles. A triangulation can be created from any planar subdivision (Figure 7.3) (Neeman, 1990).

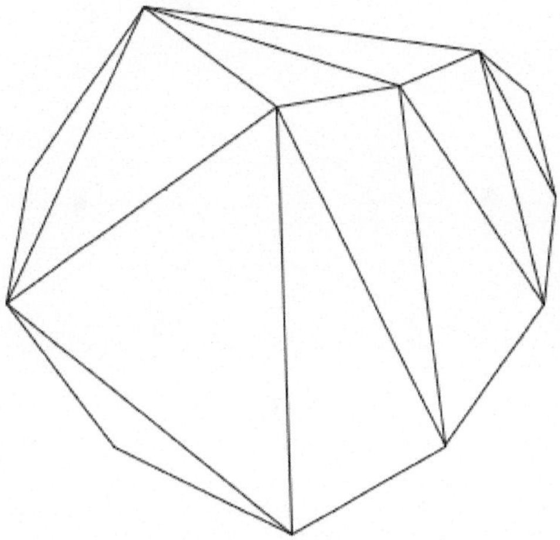

Figure 7.3. A triangulation of a point set.

Source: https://www.researchgate.net/figure/A-Triangulation-of-a-Point-Set-X-of-Size_fig1_343188422.

Polyhedra, polygons, edges, and vertices make up a subdivision in three dimensions. It's worth noting that, despite the fact that a vertex in 2D may contain any degree, we might still rank the edges to that this is close. This is difficult to achieve in 3D. Moreover, each polyhedron that makes up the subdivision is a planar subdivision in and of itself. The deconstruction into tetrahedra is the counterpart of a triangulation here (Fuchs et al., 1980; Fortune et al., 1991).

In some cases, a planar subdivision like the one depicted in Figure 7.3 could be useful. Many purposes, though, rely on subdivisions possessing specific qualities. The Voronoi diagram and the Delaunay triangulation are two of the most significant such characteristics (Fortune, 1987; Edelsbrunner and Mucke, 1988). The Voronoi diagram shows the area of space for which every location is the nearest. The Delaunay triangulation in the plane has the characteristic that its lowest angle is the greatest of all triangulations (Figure 7.4) (Edelsbrunner, 1987).

Figure 7.4. The Voronoi diagram and Delaunay triangulation.

Source: https://www.baeldung.com/cs/voronoi-diagram.

7.2.3. Towards Implementation

After establishing a solid foundation for Voronoi diagrams, Delaunay triangulations, and convex hulls, it is necessary to explore the problems inherent in putting this theory into reality. Implementations gradually followed the appearance of the aforementioned results. Graham's approach for calculating convex hulls was readily implemented as a categorizing algorithm's extension. Nevertheless, techniques for computing the two-dimensional Voronoi diagram and Delaunay triangulation, as well as the three-dimensional convex hull, encountered difficulties. Finding an adequate data structure to capture the geometry was challenging (Dwyer, 1987; Edelsbrunner, 1989). Such issues occurred when the complexity of incidence correlations increased in the following level. A vertex (or face) no longer has a defined degree. Furthermore, and one of these parameters may be controlled (as in the instance of triangulation), we never manage both (Dwyer, 1987; Dobkin and Silver, 1990).

Baumgart (1976) noticed that the edge was the optimal structure for saving and managing planar subdivisions. In specifically, each edge of a planar subdivision links and divides precisely two vertices. Baumgart created a winged edge structure that saves this topological data in an edge. Each vertex (or face) represents an ending of an edge (according to that relates to it). It's adequate to detect all adjoining edges to a vertex and all edges enclosing a face either in that direction. Figure 7.5 illustrates the fundamental winged edge structure. For every vertex-face pair, the thick edge contains data about just the vertices it connects, the faces it divides, and the nearby edges.

While the Baumgart method is a good starting point for expressing subdivisions, it requires some enhancements to make completely functional. It has a hard time showing subdivisions with gaps. It makes an unnecessary distinction among a subdivision's primal and dual representations (as described, for instance by the Voronoi diagram and Delaunay triangulation). Additionally, it lacks succinct primitives for traversing a subdivision (Figure 7.5) (Dobkin and Edelsbrunner, 1987; Dobkin et al., 1988).

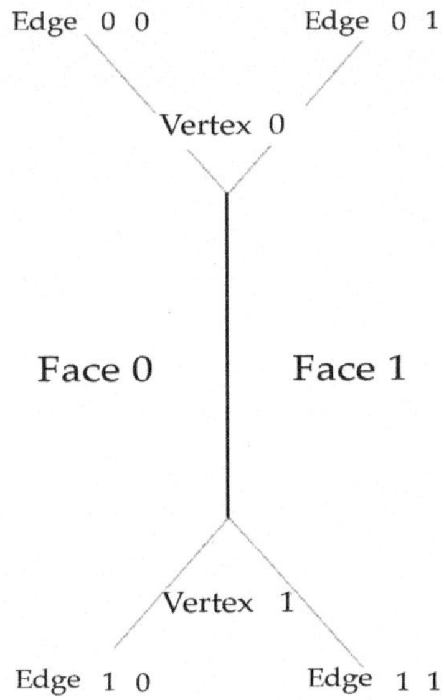

Figure 7.5. The winged edge data structure.

Source: https://www.semanticscholar.org/paper/Hierarchical-structure-to-winged-edge-structure%3A-a-Chan-Tan/ad5528d825ee32c40a8800f-9938163ce5cb04026.

For instance, the subdivision seen in Figure 7.6 (in which the dark sections represent holes) is hard to control utilizing winged edge techniques.

It is particularly challenging to follow the edges near to the central vertex. It is due to the holes provide no indication of their limits, and hence we never readily get attachable bits of the boundary (Figure 7.6).

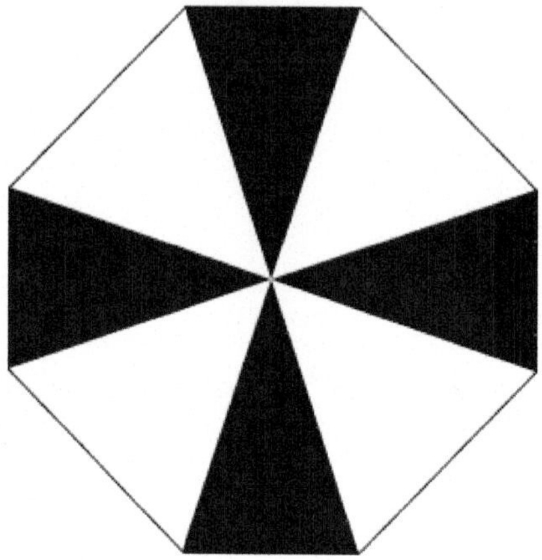

Figure 7.6. Non-winged edge structure representing the subdivision with holes.

Source: https://collaborate.princeton.edu/en/publications/computational-geometry-and-computer-graphics.

In practice, the value of a data structure is determined by the difficulty of its implementation of a Voronoi diagram method. Prior to the work described here, applications needing Voronoi diagrams were often implemented using naïve approaches (requiring O(n2) algorithms) rather than the O(n log n) algorithms discussed here.

Guibas and Stolfi (1985) extended the winged edge construction in an interesting manner. They begin by observing that an edge has the ability to fulfill four functions. It serves as a connection between two vertices and a divider between two faces. To each of these activities involves the edge in a ring of edges. These edges may extend to a vertex's neighbors or establish the border of a face. Every one of such rings is represented mathematically by a double connected listing of its edges. Every edge is identified by the ring with that it relates (Delaunay, 1934; Dey, 1991).

They then design operators that enable an edge to fulfill each of its four responsibilities. Thus, we can traverse an edge ring in search of another edge in the provided direction around one specific vertex or face. Additionally, it is likely to go from one edge ring to the next. For instance, we could begin with the edge ring equivalent to the vertex's nearby edges. Now we may go to the edge ring of neighboring edges to the face that the vertex and current edge are a member of. This procedure will be described in even more depth below. Their construction, dubbed the quad-edge, conveniently stores both the Delaunay triangulation and the Voronoi diagram (Chazelle and Edelsbrunner, 1988).

The quad edge structure's storage schema views an edge 'e' as connecting an Origin to a Destination and dividing the Left and Right bounding faces. They define the elementary operators Org, Dest, Left, and Right for the purpose of extracting these edge elements.

Edges are classified according to their equivalence classes. We have four equal representations for the winged edge seen in Figure 7.7:

- The edge between Vertex 0 and Vertex 1;
- The edge between Face 0 and Face 1;
- The edge between Vertex 1 and Vertex 0;
- The edge between Face 1 and Face 0.

Initial one of all is the edge seen in the initial illustration. The third edge is the similar as the second, but angled in the other way. The second and fourth are identical to such edges. Due to the fact that all four edges are processed identically, there is no more a distinction between primitive and dual edges. They describe the elementary operators Rot, Sym, and Flip that are used to move between these four factors (Brisson, 1989; Bern et al., 1990). These operations work in concert to make it simple to move among subdivisions and their duals. Additionally, we can investigate subdivision directly by following the edges surrounding a vertex or face. To continue our journey to different vertices or faces, we must change quad edge equivalence classes. To do this, we simply consider oriented edge rings centered on vertices and faces. For instance, the quad-edge seen in Figure 7.7 might be a component of four edge rings. It is possible to traverse such edge rings either in a positive (anticlockwise) or opposite (clockwise) way. For instance, the ring of edges surrounding Vertex 1 begins with Edge 1 0 in a positive traversal and with Edge 1 1 in a negative traversal (Bern et al., 1990; Bern and Eppstein, 1991).

7.2.4. Applications of Voronoi Diagrams and Delaunay Triangulations

In practice, the Delaunay triangulation, and the subdivisions it causes are frequently encountered. Common uses necessitate the division of a space.

The subdivision's vertices could show points where data has been tested (either computationally or physically). In terms of certain computation, the subdivision's facets could show homogeneous regions of space. Theoretical outcomes from computational geometry are frequently utilized to demonstrate that the Delaunay triangulation is the correct one (Bentley, 1975; Beichl and Sullivan, 1980). The Delaunay triangulation is then created and traversed using the algorithmic developments mentioned previously. In other cases, the Delaunay triangulation's properties are being utilized to rationalize the utilization of an easy algorithm by demonstrating that it produces the right answer. We'll go over a few scenarios where these concepts are used. See Voronoi (1991); Painter and Sloan (1989) for more uses and characteristics of Voronoi diagrams and Delaunay triangulations (Figure 7.7) (Bentley, 1975).

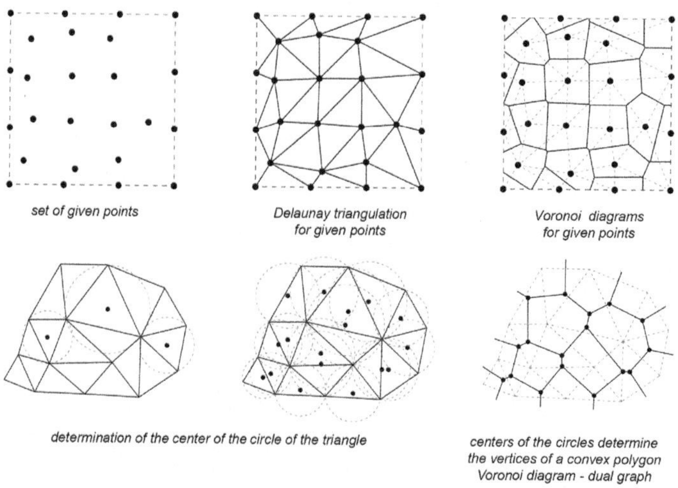

Figure 7.7. Delaunay triangulation and Voronoi diagram.

Source: https://www.researchgate.net/figure/Delaunay-Triangulation-and-Voronoi-Diagram-divisions-also-represent-a-dual-graph_fig1_311521487.

Almgren (1991) discusses the issues of determining the shortest area surface encompassing a wireframe or the lowest area surface partitioning space into given a prescription volume area. In the field of minimal surface computation, this is a basic issue. "Growing" of that kind surface areas from a soap bubble clump type geometry is a discrete estimation to the low surface issue. The primary objective is to comprehend the issue in discrete configurations and then apply what we've learned to the progress along the way. A Voronoi cell evolver is used to represent the soap bubble clusters. A computer software is used to implement the 3D Voronoi diagram. Moving controlling sites of numerous distinct colors creates and evolves combinatorial geometry (one color for every suitable cell of the suitable final soap bubble group). This geometry is being utilized to determine locations as well as to make pictures and movies of the operation utilizing computer graphics methods. Their work is based on theoretical advances in the Delaunay triangulation, and the data structure of Dobkin and Laszlo (1989) is crucial to their execution. In the subject of visualizing mathematical processes, this work is important. Without such introduction of computational geometry and graphics hardware methods, this would not have been conceivable (Barber, 1992).

The challenge of producing mesh for resolving partial differential equations is a relatable use. The calculation of inviscid transonic flows over aerodynamic structures given in Baker (1989, 1991); is an example of this scenario. The purpose is to create a triangulation of the space surrounding the aerodynamic form that is well behaved by the individual simplices (in this example, tetrahedra). Finite element techniques can yield stable solutions to the necessary equations with some tetrahedra. In this scenario, the worst-case ratio among the side lengths of a particular tetrahedron and the radius/diameter of the sphere contained in the tetrahedron is used to determine it's well behavedness. The triangulation is said to be properly behaved if this ratio develops as O(1).

A two-dimensional Delaunay triangulation or a comparable form is frequently used as the beginning point. To turn this triangulation into a 3D mesh, many approaches have been used. The most popular combines Delaunay methods with the advancing front method, that expands in a conforming way from the original triangulation. These strategies have shown to be effective as front ends for solvers that use 400,000-point sets to build 2.4-million-point meshes. On a single processor Cray 2, the triangulation was reported to take half an hour to compute.

7.3. THE BASIC TECHNIQUES IN COMPUTER GRAPHICS

The construction of realistic depictions of complicated situations is a major topic in computer graphics. This pursuit led to the study of algorithms for correctly drawing three-dimensional images on a graphic representation in the 1960s. One criterion is that invisible line or invisible surfaces be removed from a three-dimensional model. The earliest effective techniques were those that extended established paradigms to geometric domains. These developments sparked significant advances in computational geometry study. The findings of this study have led to the development of new uses of computer graphics. We investigate two solutions to the invisible surface issue and demonstrate how they might be generalized (Arvo and Kirk, 1987).

In its most basic version, invisible surface elimination analyzes a three-dimensional image made up of polygons and asks whether elements of the image appear on a plain screen (given certain illumination conditions). We may presume that all view modifications and projections have been completed for our needs. All polygons have been triangulated, giving us a set of potentially overlapping triangles in the plane wherein the observable picture is to be built. The triangles carry additional data that can be used to identify whichever of two triangles is closest to the observer (and so is viewable). The visible triangles and sections of triangles must next be identified in order to solve the invisible surface elimination challenge.

The scan line algorithm of Watkins (1970) and the space subdivision algorithm of Warnock (1991) were two early famous algorithms for this problem. Established paradigms in algorithm creation were lengthened to the geometric realm by such algorithms. Scanline algorithms are advanced outgrowths of for-loop type structures found in programming languages. Simple recursion is used in area subdivision algorithms.

The Watkins scan line method recognizes and arranges events that will happen throughout the calculation. Events occur when the collection of visible triangles alters. These modifications are caused by the emergence of new triangles or the collision of present triangles. The order is found by the y-coordinates of the events. The events are listed in ascending y-coordinate order. A data structure is monitored and upgraded among events to enable effective analysis of individual events.

The scanline processing starts with them trying to sort of all input triangles' y-coordinates. And there are three kinds of events. Initially, a triangle may be added or removed. Or else, a triangle's edges of interest may change. This happens when you come across the center (in aspects of y-coordinates) vertex. Eventually, the edges of current triangles may cross. The initial sort can determine events of the first two types. As the algorithm progresses, events of the third category must be identified. It is feasible to use it quickly since triangle edges can only connect if they correspond to triangles in our data structure that are nearby. As a result, by keeping a triangular ordering and watching variations in the order, we can discover third-type occurrences. The mathematical foundation for scan line algorithms of this sort may be traced back to Hadwiger's work (1995).

Area subdivision algorithms produce significant use of fundamental concepts. At first, the plane is seen as a simple rectangle. Furthermore, the triangles in this rectangle are treated as the set of all input triangles. The following is a recursive procedure:

- If the rectangle crosses no triangles, foreach existing rectangle.
- Return after coloring the rectangle with the background layout.
- Else if the rectangle crosses 1 triangle else.
- Return the triangle that has been clipped to the rectangle's boundaries.
- If a triangle entirely covers the rectangle(s).
- Remove any triangles that are in front of the front-most covered triangle.
- Return when that triangle has been sufficiently trimmed.
- Output that triangle suitably clipped and return.
- Else if more than one triangle exists in the rectangle, divide it into four subrectangles, identify which triangles overlap each subrectangle, and repeat.

This recursion ends when the entire image (as a collection of monochromatic triangles) is produced.

REFERENCES

1. Almgren, F., (1993). The geometric calculus of variations and modeling natural phenomena. In: Statistical Thermodynamics and Differential Geometry of Microstructured Materials (pp. 1–5). Springer, New York, NY.
2. Arvo, J., & Kirk, D., (1987). Fast ray-tracing by ray classification. ACM SIGGRAPH Computer Graphics, 21(4), 55–64.
3. Aurenhammer, F., (1991). Voronoi diagrams—A survey of a fundamental geometric data structure. ACM Computing Surveys (CSUR), 23(3), 345–405.
4. Baker, T. J., (1989). Automatic mesh generation for complex three-dimensional regions using a constrained Delaunay triangulation. Engineering with Computers, 5(3), 161–175.
5. BAKER, T., (1991). Unstructured meshes and surface fidelity for complex shapes. In: 10th Computational Fluid Dynamics Conference (pp. 1591).
6. Barber, C. B., (1992). Computational Geometry with Imprecise Data and Arithmetic (pp. 1–110). Princeton University.
7. Baumgart, B. G., (1975). A polyhedron representation for computer vision. In: Proceedings of the National Computer Conference and Exposition (pp. 589–596).
8. Beichl, I., Moseman, E., Sullivan, F., & Bowie, M., (2011). Computing network reliability coefficients. In: Proceedings of the Forty-Second Southeastern International Conference on Combinatorics, Graph Theory and Computing (Vol. 207, pp. 111–127).
9. Bentley, J. L., & Ottmann, T. A., (1979). Algorithms for reporting and counting geometric intersections. IEEE Transactions on Computers, 28(09), 643–647.
10. Bentley, J. L., (1975). Multidimensional binary search trees used for associative searching. Communications of the ACM, 18(9), 509–517.
11. Bern, M., & Eppstein, D., (1991). Polynomial-size nonobtuse triangulation of polygons. In: Proceedings of the Seventh Annual Symposium on Computational Geometry (pp. 342–350).
12. Bern, M., Dobkin, D., Eppstein, D., & Grossman, R., (1990). Visibility with a moving point of view. In: SODA (Vol. 90, pp. 107–117).
13. Bern, M., Eppstein, D., & Gilbert, J., (1994). Provably good mesh generation. Journal of Computer and System Sciences, 48(3), 384–409.
14. Brisson, E., (1993). Representing geometric structures and dimensions: Topology and order. Discrete & Computational Geometry, 9(4), 387–426.

15. Chazelle, B., & Edelsbrunner, H., (1992). An optimal algorithm for intersecting line segments in the plane. Journal of the ACM (JACM), 39(1), 1–54.
16. Davy, J. R., & Dew, P. M., (1989). A note on improving the performance of Delaunay triangulation. In: New Advances in Computer Graphics (pp. 209–226). Springer, Tokyo.
17. Delaunay, B., (1934). On the empty sphere. Izv Acad. Sciences of the USSR, Department of Mathematical and Natural Sciences,, 7(793–800), 1, 2.
18. Dey, T. K., (1991). Triangulation and CSG representation of polyhedra with arbitrary genus. In: Proceedings of the Seventh Annual Symposium on Computational Geometry (pp. 364–371).
19. Dobkin, D. P., & Edelsbrunner, H., (1987). Space searching for intersecting objects. Journal of Algorithms, 8(3), 348–361.
20. Dobkin, D. P., & Laszlo, M. J., (1987). Primitives for the manipulation of three-dimensional subdivisions. In: Proceedings of the Third Annual Symposium on Computational Geometry (pp. 86–99).
21. Dobkin, D., & Lipton, R. J., (1974). On some generalizations of binary search. In: Proceedings of the sixth annual ACM Symposium on Theory of Computing (pp. 310–316).
22. Dobkin, D., & Lipton, R. J., (1976). Multidimensional searching problems. SIAM Journal on Computing, 5(2), 181–186.
23. Dobkin, D., & Silver, D., (1990). Applied computational geometry: Towards robust solutions of basic problems. Journal of Computer and System Sciences, 40(1), 70–87.
24. Dobkin, D., Guibas, L., Hershberger, J., & Snoeyink, J., (1988). An efficient algorithm for finding the CSG representation of a simple polygon. In: Proceedings of the 15th Annual Conference on Computer Graphics and Interactive Techniques (pp. 31–40).
25. Dwyer, R. A., (1987). A faster divide-and-conquer algorithm for constructing Delaunay triangulations. Algorithmica, 2(1), 137–151.
26. Edelsbrunner, H., & Mücke, E. P., (1988). Simulation of simplicity. In: Proc. ACM Symposium on Computational Geometry (pp. 118–133). Urbana-Champaign, Illinois.
27. Edelsbrunner, H., (1987). Algorithms in Combinatorial Geometry (Vol. 10). Springer Science & Business Media.
28. Edelsbrunner, H., Kirkpatrick, D., & Seidel, R., (1983). On the shape of a set of points in the plane. IEEE Transactions on Information Theory, 29(4), 551–559.

29. Fortune, S., & Milenkovic, V., (1991). Numerical stability of algorithms for line arrangements. In: Proceedings of the Seventh Annual Symposium on Computational Geometry (pp. 334–341).
30. Fortune, S., (1987). A sweeping algorithm for Voronoi diagrams. Algorithmica, 2(1), 153–174.
31. Fuchs, H., Kedem, Z. M., & Naylor, B. F., (1980). On visible surface generation by a priori tree structures. In: Proceedings of the 7th Annual Conference on Computer Graphics and Interactive Techniques (pp. 124–133).
32. Graham, R. L., (1972). An efficient algorithm for determining the convex hull of a finite planar set. Info. Pro. Lett., 1, 132, 133.
33. Graham, R. W., & Mead, J. I., (1987). Environmental Fluctuations and Evolution of Mammalian Faunas During the Last Deglaciation in North America (pp. 10–15).
34. Guibas, L., & Stolfi, J., (1985). Primitives for the manipulation of general subdivisions and the computation of Voronoi. ACM Transactions on Graphics (TOG), 4(2), 74–123.
35. Hadwiger, H., (1955). Euler's characteristic and combinatorial geometry,101–110.
36. Haussler, D., Littlestone, N., & Warmuth, M. K., (1988). Predicting [0, 1]-functions on randomly drawn points. In: Annual Workshop on Computational Learning Theory: Proceedings of the First Annual Workshop on Computational Learning Theory (Vol. 3, No. 05, pp. 280–296).
37. Knuth, D. E., (1973). The art of computer programming. Sorting and Searching, 3, Ch-6.
38. Laszlo, M. J., (1987). A Data Structure for Manipulating Three-Dimensional Subdivisions (pp. 1–15). Princeton University.
39. Lienhardt, P., (1989). Subdivisions of n-dimensional spaces and n-dimensional generalized maps. In: Proceedings of the Fifth Annual Symposium on Computational Geometry (pp. 228–236).
40. Max, N., Hanrahan, P., & Crawfis, R., (1990). Area and volume coherence for efficient visualization of 3D scalar functions. In: Proceedings of the 1990 Workshop on Volume Visualization (pp. 27–33).
41. Neeman, H., (1990). A decomposition algorithm for visualizing irregular grids. ACM SIGGRAPH Computer Graphics, 24(5), 49–56.
42. Painter, J., & Sloan, K., (1989). Antialiased ray-tracing by adaptive progressive refinement. In: Proceedings of the 16th Annual Conference on Computer Graphics and Interactive Techniques (pp. 281–288).

43. Palios, L., (1992). Decomposition Problems in Computational Geometry (pp. 5–7). Doctoral dissertation, Princeton University.
44. Paterson, M. S., & Yao, F. F., (1990). Efficient binary space partitions for hidden-surface removal and solid modeling. Discrete & Computational Geometry, 5(5), 485–503.
45. Peterson, D. P., (1984). Halfspace Representations of Extrusions, Solids of Revolution, and Pyramids (No. SAND-84-0572) (pp. 15–20). Sandia National Labs., Albuquerque, NM (USA).
46. Salesin, D., Stolfi, J., & Guibas, L., (1989). Epsilon geometry: Building robust algorithms from imprecise computations. In: Proceedings of the Fifth Annual Symposium on Computational Geometry (pp. 208–217).
47. Samet, H., (1990). Applications of Spatial Data Structures: Computer Graphics, Image Processing, and GIS. (pp. 35–45). Addison-Wesley Longman Publishing Co., Inc.
48. Samet, H., (1990). The Design and Analysis of Spatial Data Structures (Vol. 85, p. 87). Reading, MA: Addison-Wesley.
49. Shamos, M. I., & Hoey, D., (1975). Closest-point problems. In: 16th Annual Symposium on Foundations of Computer Science (sfcs 1975) (pp. 151–162). IEEE.
50. Shamos, M. I., & Hoey, D., (1976). Geometric intersection problems. In: 17th Annual Symposium on Foundations of Computer Science (sfcs 1976) (pp. 208–215). IEEE.
51. Shamos, M. I., (1975). Geometric complexity. In: Proceedings of the Seventh Annual ACM Symposium on Theory of Computing (pp. 224–233).
52. Sproull, R. F., (1991). Refinements to nearest-neighbor searching ink-dimensional trees. Algorithmica, 6(1), 579–589.
53. Warnock, J. E., (1969). A Hidden Surface Algorithm for Computer Generated Halftone Pictures (pp. 15–25). The University of Utah.
54. Watkins, G. S., (1970). A Real-Time Visible Surface Algorithm (pp. 35–45). University of Utah. Computer Science Department.
55. Wilhelms, J., & Van, G. A., (1990). Octrees for faster isosurface generation. ACM SIGGRAPH Computer Graphics, 24(5), 57–62.
56. Yao, F. F., (1980). Efficient dynamic programming using quadrangle inequalities. In: Proceedings of the Twelfth Annual ACM Symposium on Theory of Computing (pp. 429–435).

CHAPTER

8 Computational Geometry in Graph Drawing

Contents

8.1. INTRODUCTION

Graph drawing is an important component of the developing subject of information visualization, and it addresses the difficulty of building geometric representations of graphs.

Information retrieval, database systems, software engineering, computer-aided design, and visual interfaces are all examples of where graph drawing may be used. Graph drawing research has been done in a variety of disciplines, including human-computer interaction (HCI) (visual languages,

Graphical user interfaces, software visualization), order theory), algorithmics (graph algorithms, data structures, computational geometry, VLSI), and discrete mathematics (topological graph theory, geometric graph theory (Figure 8.1) (Asano et al., 1985; Di Battista et al., 1997).

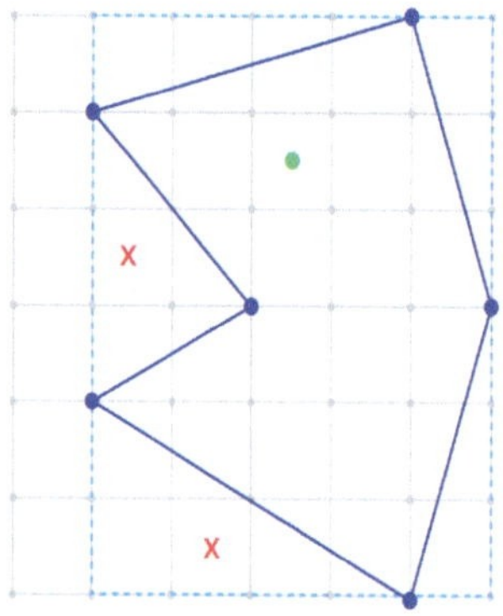

Figure 8.1. Computational geometry drawing in python.

Source: https://www.toptal.com/python/computational-geometry-in-python-from-theory-to-implementation.

8.2. TYPES OF GRAPHS

To begin, we will establish basic graph terms that will be useful while creating graphs. The numbers 'n' and 'm' will be used throughout this chapter to denote the number of edges and graph vertices, and 'd' will represent the maximum vertex degree (Figure 8.2) (Cardinal, 2015).

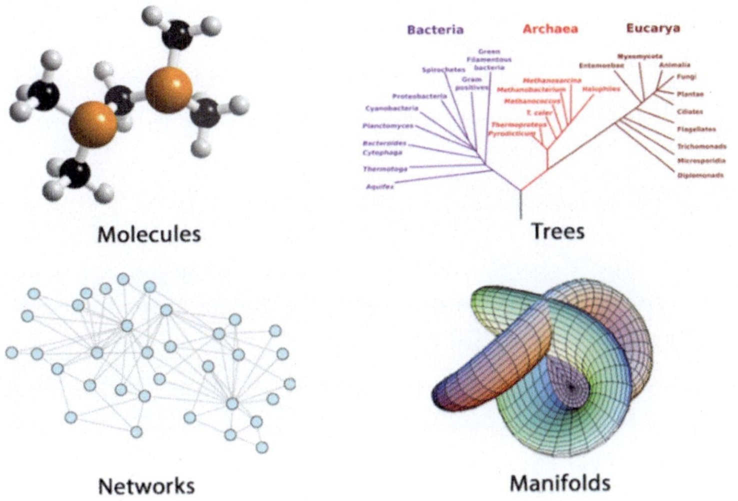

Figure 8.2. Different types of graphs showing various forms of atoms and molecules.

Source: https://flawnsontong.medium.com/what-is-geometric-deep-learning-b2adb662d91d.

1. **Degree-k Graph:** Graph with the highest degreed < k.
2. **Digraph:** Directed graph.
3. **Acyclic Digraph:** Exclusive of directed cycles.
4. **Transitive Edge:** A digraph's edge (u, v) is transitive if there is a directed path from 'u' to 'v' that does not include edge (u, v).
5. **Reduced Digraph:** Not including transitive edges.
6. **Source:** Peak of a digraph devoid of incoming edges.
7. **Sink:** Peak of a digraph devoid of outgoing edges.
8. **St-Digraph:** An acyclic digraph having one source and one drain connected by an edge.
9. **Connected Graph:** A route connects any two vertices.
10. **Biconnected Graph:** Two vertex-disjoint pathways connect any two vertices.
11. **Triconnected Graph:** Three vertex-disjoint pathways connect any two vertices.

12. **Tree:** Graph with no cycles that are connected.
13. **Rooted Tree:** Each vertex in a directed tree has a distinct vertex, the root, and each vertex is on a directed route to the root.
14. **Binary Tree:** Rooted tree where each vertex has at most two incoming edges.
15. **Layered (Di)graph:** The vertices are divided into layers, which are groups of vertices. A rooted tree may be thought of as a layered digraph, with each layer consisting of a collection of vertices that are simultaneously distant from the root (Tamassia, 2000; Eppstein, 2009).
16. **k-Layered (Di)graph:** Layered (di)graph with 'k' layers.

8.3. TYPES OF DRAWINGS

Vertices are indicated by points (or by geometric forms such as circles or rectangles), and edges are depicted by curves, with any two edges intersecting at most once among a limited number of points in a graph drawn in this manner. With the exception of Section 4.8, which deals with 3D drawings, we are only concerned with drawings in the plane. In addition, we do not take up the issue of labeling the edges and vertices of the graph (Figure 8.3) (Di Battista et al., 1994; Tamassia, 1996).

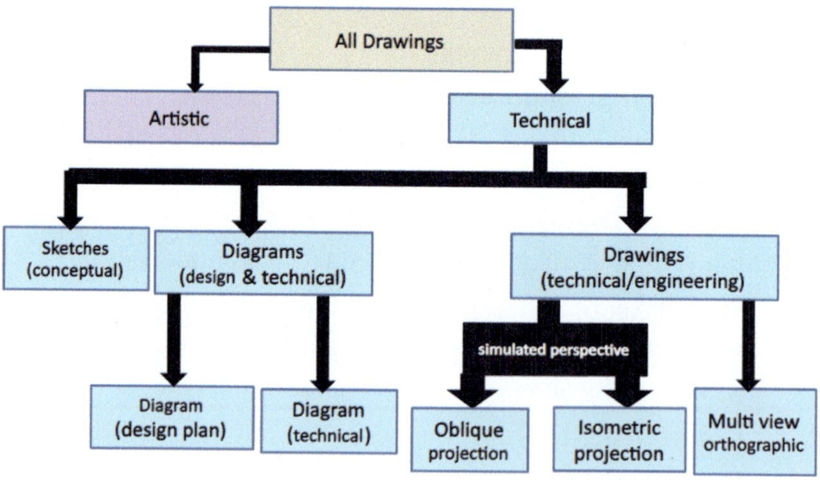

Figure 8.3. Different types of drawings.

Source: https://slideplayer.com/slide/8163491/.

1. **Polyline Drawing:** Each of the edges is made up of a polygonal chain (Figure 8.4(a)).
2. **Straight-Line Drawing:** Each edge is a section of a straight line (Figure 8.4(b)).
3. **Orthogonal Drawing:** Each of the edges is made up of a series of vertical and horizontal segments (Figure 8.4(c)).
4. **Bend:** The place where two segments of the same edge meet in a polyline drawing (Figure 8.4(a)).
5. **Crossing:** The point where two edges meet (Figure 8.4(b)).
6. **Grid Drawing:** Integer coordinates are used for vertices, crossings, and bends in a polyline drawing.
7. **Planar Drawing:** There are no edges that cross each other (see Figure 8.4(d)).
8. **Planar (Di)graph:** Acknowledges a planar drawing.
9. **Embedded (Di)graph:** Planar (di)graph having a topologic inserting (i.e., a collection of faces) that must be kept in the drawing.
10. **Upward Drawing:** Illustration of a digraph with each edge nondecreasing monotonically in the vertical way.

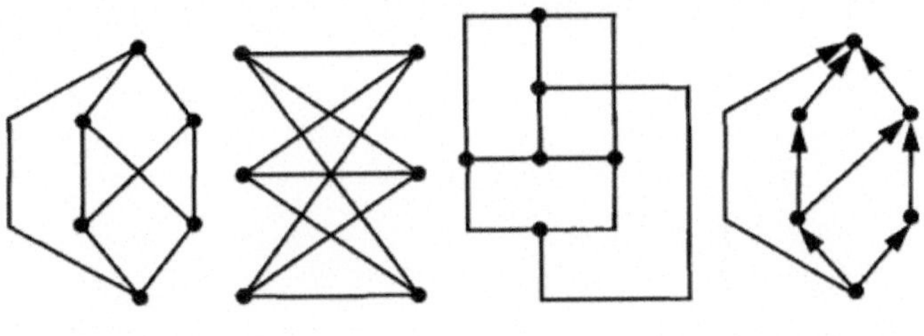

(a) (b) (c) (d)

Figure 8.4. Polyline drawings, straight-line drawings, orthogonal drawings, and planar upward drawings of an acyclic digraph are all examples of drawings (Kuchment, 2008; Agryzkov et al., 2017).

11. **Upward Planar Digraph:** Acknowledges an upward planar drawing.
12. **Layered Drawing:** Drawing a covered graph with vertices from the same layer all lying on a similar horizontal line.
13. **Face:** The unbounded portion of a plane specified by a planar drawing is known as the external face.

14. **Convex Drawing:** Each face's border is a convex polygon in this planar straight-line illustration.
15. **Visibility Drawing:** A graph is created depending on a geometric closeness relationship. A tree can be depicted as the Euclidean minimal spanning tree of a cluster of points, for example.
16. **Dominance Drawing:** The illustration of an acyclic digraph upward in such a way that there is a directed route from vertex 'u' to vertex 'V' if and only if x(u) is greater than x(v) and y(u) is greater than y(v), where the coordinates of a vertex are denoted by the letters x(O and y(.).
17. **hv-Drawing:** The illustration of a binary tree from the top down in orthogonal straight lines, with the illustrations of the subtrees of every node divided by a vertical or horizontal line.

Polyline drawings include orthogonal drawings and straight-line, which are both special examples of polyline drawings (Jeyalakshmi and Meena, 2021). Because they can approach drawings along with curved edges, polyline drawings provide a great deal of versatility in terms of design.

Edges that have more than two or three bends, on the other hand, may be difficult for the eye to "follow." Furthermore, a system that allows for the editing of polyline drawings is more complex than a system that just allows for the editing of straight-line drawings.

In this case, depending on the application, it may be preferable to use polyline or straight-line drawings. Graphs with maximum vertex degree 4 are the only ones that have orthogonal drawings if the vertices are depicted by points (Kozae et al., 2019).

8.4. PROPERTIES OF DRAWINGS

1. **Crossings (X):** In a drawing, the total number of crossings is calculated.
2. **Area:** of the drawing's convex hull.
3. **Total Edge Length:** The sum of the edge lengths.
4. **Maximum Edge Length:** An edge's maximum length.
5. **The Total Number of Bends:** This on a polyline drawing's edges.
6. **Maximum Number of Bends:** This on the edge of a polyline drawing.

7. **Angular Resolution (p):** In a polyline drawing, the smallest angle is generated by two edges, or segments of edges, incidental on the same vertex or bend.
8. **Aspect-Ratio:** The length of the longest side of the shortest rectangle with vertical and horizontal sides encompassing the drawing divided by the length of the shortest side.

There are several features that we want to consider while drawing a graph in order to do so efficiently and effectively. In visualization applications, for example, planarity, and the presentation of symmetries are highly desirable characteristics. The practice of sketching trees and acyclic digraphs from the top down is also common among artists. In general, it is necessary to keep the area of the drawing as small as possible in order to prevent wasting valuable paper space or computer screen real estate (Alon, 1981; Deussen et al., 2000). The following are examples of commonly used graph sketching esthetics:

- The overall number of crossings between edges is kept to a minimum. We'd prefer to have a flat picture, but not every graph allows for that.
- The drawing's surface area is minimized. This appeal is only significant if the drawing standard used prohibits drawings from being scaled down arbitrarily.
- The overall number of bends around the edges is kept to a minimum. This esthetic is particularly critical for orthogonal illustrations, whereas straight-line drawings just satisfy it.
- The drawing's aspect ratio is kept as small as possible. In an ideal world, we'd want to have a unit aspect ratio.
- The drawing's axial and rotational symmetries should be maximized. This esthetic may be defined even further by using a mathematical model of symmetry in graphs and pictures.

Graph optimization difficulties are inextricably linked to esthetics. However, the majority of such issues are computationally difficult. As a result, several heuristic graph drawing approaches have been created that function well in practice despite the fact that they do not always optimize esthetics (Liotta and Battista, 1995; Rebolj and Menzel, 2004).

8.5. BOUNDS

Many upper and lower bounds for certain drawing qualities have been identified for many classes of drawing kinds and graphs, including both universal and existential upper and lower bounds. Such limitations are frequently characterized by tradeoffs among different drawing qualities (Gigus and Malik, 1988; Kimoto et al., 2005). A common bond abounds that relates to all graphs of a particular class. It is possible to have an unlimited number of graphs in a class, which means that there is an endless family of graphs with the lower bound if you use existential bounds (Figure 8.5) (Giacomo et al., 2004; Chen and Kamara, 2011).

Figure 8.5. Boundaries are an integral part of a drawing.

Source https://twitter.com/cartoonconnie/status/1092907172659412996.

8.6. TRADEOFF BETWEEN AREA AND ASPECT-RATIO

Even when designing graphs with a simple structure, like trees, a number of tradeoffs for the aspect ratio and area are necessary to consider. For various sorts of tree drawings, Table 8.1 outlines selected universal constraints that can be reached concurrently on the aspect ratio and area of different kinds of drawings (Figure 8.6) (Bertolazzi et al., 2000; Samee et al., 2008).

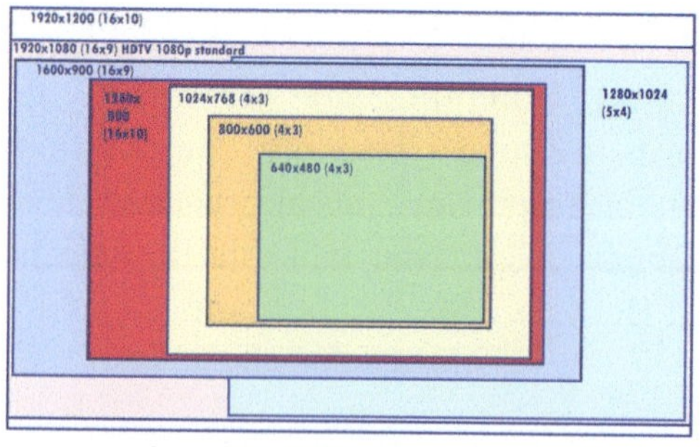

Figure 8.6. Same aspect ratio with different areas of the screen.

Source: https://www.vectornator.io/design-tips/what-is-the-aspect-ratio.

Downward planar illustrations are by far the most normal way of picturing rooted trees since they are the most realistic. With the exception of binary trees, the current straight-line illustration approaches are insufficient when it comes to meeting either the decision or the aspect ratio of the drawing. If we consider linear area, polyline drawings may be accomplished with a predetermined $O(n^a)$ aspect ratio, where an is a subjective quantity such that 0 a 1 is true (Aloupis et al., 2002; Conitzer et al., 2006).

Table 8.1. Upper Boundaries for the Size and Aspect Ratio of Tree Drawings That can be Attained Concurrently*

Class	**Drawing Type**	**Area**	**Aspect-Ratio**
Rooted tree	Layered (upward planar straight-line) grid	$O(n^2)$	O(1)
Rooted tree	Upward planar straight-line grid	O(n log n)	O(n/log n)
Binary tree	Upward planar straight-line orthogonal grid	O(n log log n)	O(1)
Rooted degree-$O(n^a)$ tree	Upward planar polyline grid	O(n)	$O(n^a)$
Binary tree	Upward planar orthogonal grid	O(n log log n)	O(n log log n/ $\log^2$ n)
Binary tree	Planar straight-line orthogonal grid	O(n log n)	O(1)
Degree-4 tree	Planar orthogonal grid	O(n)	O(1)

*We denote by an arbitrary constant such that $0 \leq a < 1$.

The area and aspect ratio of planar straight-line orthogonal grid illustrations of binary trees is O(n log log n) and 0(1), respectively, for non-upward drawings of trees, and the area and aspect ratio of planar orthogonal grid drawings of degree-4 trees are O(n log log n) and 0(1) for upward illustrations of trees. The latter technique, on the other hand, does not appear to produce visually beautiful graphics and appears to be more suited for VLSI design than for visualization applications (Pratt and Yao, 1973; Buchin et al., 2017).

8.6.1. The Tradeoff Between Angular Resolution and Area

The area and the angular resolution both have tradeoffs. Table 8.2 lists some of the universal boundaries that can be attained at the same time. Only planar polyline drawings and nonplanar straight-line drawings can obtain good simultaneous limits. Indeed, every planar straight-line illustration with optimum S_2 (1) angular resolution requires exponential area in any O(l)-degree planar network. In planar straight-line drawings, this finding may be complete to a continuous tradeoff between angular resolution and area (Dietzfelbinger et al., 1994; Zhongke et al., 2004).

Table 8.2. Asymptotic Constraints for the Angular Resolution and Area That may be Attained in Graph Drawings at the Same Time*

Class	Drawing Type	Area	Angular Resol.
Planar graph	Straight-line	$O(d^6n)$	
Planar graph	Straight-line	$O(d^3n)$	
Planar graph	Planar straight-line grid	$O(n^2)$	
Degree-O(1)	Planar straight-line	(c^n)	
Planar graph	Planar straight-line	$(c^{\rho n})$	
Planar graph	Planar straight-line	$O(b^n)$	
Planar graph	Planar polyline grid	$O(n^2)$	

*We represent by 'b' and 'c' constants such that $b > 1$ and $c > 1$.

8.6.2. Bounds on the Computational Difficulty

Embedding issues and planarity testing have been studied for their computational complexity, and the results are given in Table 8.3. It is noteworthy to note that situations that appear to be quite similar in their temporal complexity turn out to be extremely different. While planarity testing may be completed in linear time, upward planarity testing is an NP-hard problem (Mahmoodi et al., 2020). Note that upward planarity testing may be accomplished in polynomial time for some classes of digraphs if the classes are constrained.

It takes $O(n + m)$ time to produce orthogonal drawings of graphs using the approach described in Gál et al. (2012); Kennedy et al. (2014). The computational cost of various broad ways for drawing graphs is difficult to quantify since each of the phases in a particular methodology may be performed using a variety of different algorithmic frameworks and approaches. Symmetries in drawings of generic graphs are difficult to detect and depict (Daniel et al., 2007; Pizzarelli et al., 2013) because they are NP-complete.

Table 8.3. Planarity Testing and Encoding of Time Complexity of Chosen Graph Drawing Issues

Problem	**Class**	**Complexity**
Minimize crossings	General graph	NP-hard
Maximum planar subgraph Planarity testing and computing a planar embedding Maximal planar subgraph	General graph General graph General graph	XP-hard
Minimize crossings in a layered drawing	2-layered graph with preassigned order on one layer	NP-hard
Upward planarity testing	General digraph	NP-hard
Upward planarity testing	Embedded outerplanar digraph	–
Upward planarity testing	Outerplanar digraph	–
Upward planarity testing	Single-source digraph	(9(n)

The computational intricacy of numerous challenges arising in the construction of polyline and straight-line representations of planar graphs is summarized in Table 8.4. Planar straight-line illustrations with the defined maximum angular resolution, regulated angles, or edge lengths are NP-hard to create.

For degree-4 graphs, creating orthogonal grid illustrations with O(n) bends and O(n) area may be done in O(n) time. Planar orthogonal grid illustrations with similar restrictions on the area and number of bends may be created in O(n) time if the input graph is planar. It's NP-hard to reduce the number of bends in planar orthogonal drawings. However, if the input graph has a predefined encapsulation or has a degree of at most 3, the issue can be handled in polynomial time (Maass, 1997; de Assis et al., 2020).

The area of a two-dimensional straight-line upward layered depiction of a tree that depicts isomorphisms and symmetries of subtrees is minimized using a polynomial-time technique centered on linear programming (Traub and Woźniakowski, 1976). It's worth noting that the drawing created by this method isn't a grid drawing in general.

The area of hv-drawings of binary trees is minimized using a polynomial-time technique based on dynamic programming. Many additional drawing optimization problems, such as edge-length and area reduction in planar grid drawings, are NP-hard for trees. Determining if a tree can be rendered as the Euclidean minimal spanning tree of a collection of points in the plane is also NP-hard (Compton and Henson, 1990; Freivalds and Karpinski, 1994).

Rooted trees provide efficient drawing methods that ensure strong universal upper limits on the area.

Table 8.4. The Temporal Complexity of Chosen Graph Drawing Problems: Planar Graphs Drawn in Straight Lines and Polylines

Problem	**Class**	**Complexity**
Drawing of a planar straight line with prescribed angles	Embedded planar graph	NP-hard
Drawing of a planar straight line with specified angles	Maximal planar graph	@(n)
Planar straight-line illustration with maximum angular resolution	Planar graph	NP-hard
Two-dimensional polyline grid drawing with O(n) area, O(n) bends, and O(l/d) angular resolution	Planar graph	$\Theta(n)$
Two-dimensional straight-line grid drawing with O(n) area and $S_2(1/n^2)$ angular resolution	Planar graph	@(n)
Planar straight-line convex grid	Triconnected planar	$\Theta(n)$
Visibility symbol with O(n) area	Planar graph	$\Theta(n)$
Planar straight-line strictly con-	Triconnected planar	$\Theta(n)$
Drawing with $O(n^2)$ area and $S_2(1/n^2)$ angular resolution	Graph	–
Dominance drawing with a minimum area	St-digraph	–
Vex drawing	Graph	–
Planar straight-line symmetric drawing	Outerplanar graph	@(n)
Upward planar grid straight-line	Reduced planar	@(n)
Drawing with O(n) area and O(n) bends	Digraph	–
Upward planar polyline grid	Upward planar	@(n)

8.7. METHODS FOR DRAWING GRAPHS

In this part, we'll show you how to make graph drawings using several strategies.

8.7.1. Force-Directed Approach

Force-directed approaches use a physical model in which the graph's vertices and edges are considered as physical objects subject to multiple forces to create straight-line drawings of generic graphs. The graph progresses from an initial random drawing to a final drawing, which is the physical system's local lowest energy configuration. In the literature, force-directed methods are also known as physical simulation methods and spring methods (Table 8.5) (Chen, 1976; Chen et al., 2006).

Table 8.5. The Time Complexity of Selected Graph Drawing Problems: Orthogonal Drawings of Graphs with a Degree at most 4

Class	Problem	Complexity	References
Planer degree-3 graph	with O(n) bends and $O(n^2)$ area	O(n), $O(n^5$ log n)	Chen, 1976
Two-dimensional degree-4 graph	Drawing of a two-dimensional orthogonal grid with O(n) area and O(n) bends	@(n)	Chen, 1976 Traub and Woźniakowski, 1976
Two-dimensional degree-4 graph	It is a drawing of a two-dimensional orthogonal grid with the fewest bends possible.	NP-hard	–
Two-dimensional degree-3 graph	Drawing of a two-dimensional orthogonal grid with the fewest bends and $O(n^2)$ area.	$S_2(n)$, $O(n^5$ log n)	Chen, 1976
Embedded two-dimensional	Two-dimensional orthogonal grid	O(n), $001^{7/4}$ log n)	Chen, 1976, Traub and Woźniakowski, 1976
Degree-4 graph	Drawing with the least number of bends and $O(n^2)$ area	–	–

Every pair of vertices in Eades' spring embedder is linked by a "spring.":

Rather than solving a set of differential equations, numerical techniques are frequently used to simulate the system's development. The simulation is made up of a series of phases (see Figure 8.7). The resultant force 'f' on each vertex 'v' is computed at each step, and 'v' is translated by A f, where A is a dislocation factor (Table 8.6). Aldridge et al. (2014) pioneered the force-directed technique. Among the notable developments are. Specifically:

- The technique of uses forces to put vertices so that their Euclidean distance in the illustration is equivalent to their graph-theoretic distance d(u, v) in the graph; consequently, edge (u, v) has ideal length cl in their approach (u, v) (Fich et al., 1985).

Table 8.6. Trees: The Temporal Complexity of Chosen Graph Drawing Issues*

Class	Problem	Complexity
Degree-4 tree	Minimize area in planar orthogonal grid drawing	NP-hard
Degree-4 tree	In a flat orthogonal grid drawing, reduce the total/maximum edge length.	NP-hard
Rooted tree	Reduce the size of a planar straight-line upward stacked grid illustration that shows subtree symmetries and isomorphisms	NP-hard
Binary tree	Minimize area in hv-drawing	Q (n), O(n n log n)
Rooted tree	Reduce the size of a planar straight-line upward covered graphic that shows subtree isomorphisms and symmetries	Q(n), $O(n^k)$
Rooted tree	Planar polyline upward grid drawing with O(n) area	@(n)
Rooted tree	Planar straight-line upward layered grid drawing with O(n) area	$\Theta(n)$
Rooted tree	Planar straight-line upward grid drawing with O(n log n) area	$\Theta(n)$

*By A, we mean: a constant such that $k \geq 1$.

- The scale of the drawing is regulated by presuming that the boundary of the predefined drawing region operates as a "wall" in, which proposes a complicated system of forces akin to those of subatomic elements and astronomical bodies (Hromkovič, 1999).
- The approach of does not directly employ forces but instead constructs an energy function that considers edge crossings, edge lengths, and vertex distribution and then uses the simulated

annealing method to create a drawing with low energy (Pirandola and Lloyd, 2008).

It's worth noting that Tutte's barycentric approach, a famous drawing algorithm for planar graphs, may be recast as a restricted force-directed method (Figure 8.7).

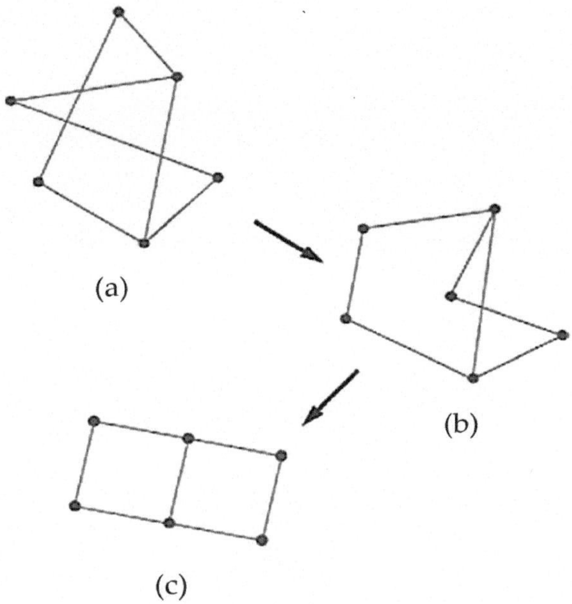

Figure 8.7. Model of the development of a drawing in the spring embedder algorithm: (a) initial drawing; (b) drawing at an intermediary step; (c) final drawing.

Drawing algorithms depending on the force-directed technique provide significant benefits, according to an experimental comparison of several force-directed approaches. They are, in particular:

- comparatively simple to execute;
- easy to parameterize;
- easy to expand by adding new forces;
- for short graphs with a regular structure, this method is typically effective;
- can frequently find and present symmetries in graphs;
- capable of retaining the user's "mental map" by allowing the drawing to evolve from its original state to its final state.

Disadvantages include:

- Large graphs take a long time to load and execute;
- Only drawings with straight lines are accepted;
- The theoretical analysis of the produced drawings' attributes is challenging.

8.7.2. Layering Approach

For producing polyline drawings of directed graphs, the layering technique turns the diagram into a covered digraph and then builds a layered drawing. The following are the major phases of a typical optimization method on the layering technique (Figure 8.8):

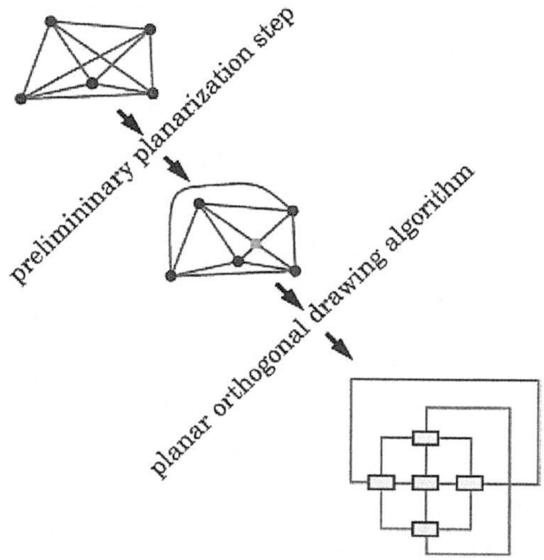

Figure 8.8. The planarization method is depicted in a diagram.

- Add each vertex to a layer in order to maximize the number of upward-facing edges;
- Replace the real vertices along the edges that cross layers with fake vertices, like, each edge in the resultant digraph joins vertices in subsequent layers;
- Drastic the vertices on every layer to reduce the number of crossings;
- Alter the vertices' positions in each layer with the purpose of uniformly spreading the vertices and reducing the number of bends.

As a result, heuristics must be used to the majority of the subproblems that arise during the different phases in the process. Pardalos and Kovoor (1990) is credited with developing the layering method. The most significant advancements in this method are documented.

REFERENCES

1. Agryzkov, T., Oliver, J. L., Tortosa, L., & Vicent, J. F., (2017). Different Types of Graphs to Model a City. WIT Trans. Eng. Sci., 118, 71–82.
2. Aldridge, M., Baldassini, L., & Johnson, O., (2014). Group testing algorithms: Bounds and simulations. IEEE Transactions on Information Theory, 60(6), 3671–3687.
3. Alon, N., (1981). On the number of subgraphs of prescribed type of graphs with a given number of edges. Israel Journal of Mathematics, 38(1), 116–130.
4. Aloupis, G., Cortés, C., Gómez, F., Soss, M., & Toussaint, G., (2002). Lower bounds for computing statistical depth. Computational Statistics & Data Analysis, 40(2), 223–229.
5. Asano, T., Edahiro, M., Hiroshi, I. M. A. I., Masao, I. R. I., & Murota, K., (1985). Practical use of bucketing techniques in computational geometry. In: Machine Intelligence and Pattern Recognition (Vol. 2, pp. 153–195). North-Holland.
6. Bertolazzi, P., Di Battista, G., & Didimo, W., (2000). Computing orthogonal drawings with the minimum number of bends. IEEE Transactions on Computers, 49(8), 826–840.
7. Buchin, K., Buchin, M., Meulemans, W., & Mulzer, W., (2017). Four soviets walk the dog: Improved bounds for computing the Fréchet distance. Discrete & Computational Geometry, 58(1), 180–216.
8. Cardinal, J., (2015). Computational geometry column 62. ACM SIGACT News, 46(4), 69–78.
9. Chen, C. H., (1976). On information and distance measures, error bounds, and feature selection. Information Sciences, 10(2), 159–173.
10. Chen, J., Huang, X., Kanj, I. A., & Xia, G., (2006). Strong computational lower bounds via parameterized complexity. Journal of Computer and System Sciences, 72(8), 1346–1367.
11. Chen, Y., & Kamara, J. M., (2011). A framework for using mobile computing for information management on construction sites. Automation in Construction, 20(7), 776–788.
12. Compton, K. J., & Henson, C. W., (1990). A uniform method for proving lower bounds on the computational complexity of logical theories. Annals of Pure and Applied Logic, 48(1), 1–79.
13. Conitzer, V., Davenport, A., & Kalagnanam, J., (2006). Improved bounds for computing Kemeny rankings. In: AAAI (Vol. 6, pp. 620–626).

14. Daniel, J. H., Sawant, A., Teepe, M., Shih, C., Street, R. A., & Antonuk, L. E., (2007). Fabrication of high aspect-ratio polymer microstructures for large-area electronic portal x-ray imagers. Sensors and Actuators A: Physical, 140(2), 185–193.
15. De Assis, T. A., Dall'Agnol, F. F., & Cahay, M., (2020). Tradeoff between proximity and aspect-ratio in optimizing the field enhancement factor of large-area field emitters. In: 2020 33rd International Vacuum Nanoelectronics Conference (IVNC) (pp. 1, 2). IEEE.
16. Deussen, O., Hiller, S., Van, O. C., & Strothotte, T., (2000). Floating points: A method for computing stipple drawings. In: Computer Graphics Forum (Vol. 19, No. 3, pp. 41–50). Oxford, UK and Boston, USA: Blackwell Publishers Ltd.
17. Di Battista, G., Eades, P., Tamassia, R., & Tollis, I. G., (1994). Algorithms for drawing graphs: An annotated bibliography. Computational Geometry, 4(5), 235–282.
18. Di Battista, G., Garg, A., Liotta, G., Tamassia, R., Tassinari, E., & Vargiu, F., (1997). An experimental comparison of four graph drawing algorithms. Computational Geometry, 7(5, 6), 303–325.
19. Dietzfelbinger, M., Kutyłowski, M., & Reischuk, R., (1994). Exact lower time bounds for computing Boolean functions on CREW PRAMs. Journal of Computer and System Sciences, 48(2), 231–254.
20. Eppstein, D., (2009). Graph-theoretic solutions to computational geometry problems. In: International Workshop on Graph-Theoretic Concepts in Computer Science (pp. 1–16). Springer, Berlin, Heidelberg.
21. Fich, F. E., Meyer, A. D. H. F., Ragde, P., & Wigderson, A., (1985). One, two, three.. infinity: Lower bounds for parallel computation. In: Proceedings of the Seventeenth Annual ACM Symposium on Theory of Computing (pp. 48–58).
22. Freivalds, R., & Karpinski, M., (1994). Lower space bounds for randomized computation. In: International Colloquium on Automata, Languages, and Programming (pp. 580–592). Springer, Berlin, Heidelberg.
23. Gál, A., Hansen, K. A., Koucký, M., Pudlák, P., & Viola, E., (2012). Tight bounds on computing error-correcting codes by bounded-depth circuits with arbitrary gates. In: Proceedings of the Forty-Fourth Annual ACM Symposium on Theory of Computing (pp. 479–494).
24. Giacomo, E. D., Didimo, W., Liotta, G., & Meijer, H., (2004). Computing radial drawings on the minimum number of circles. In: International Symposium on Graph Drawing (pp. 251–261). Springer, Berlin, Heidelberg.

25. Gigus, Z., & Malik, J., (1988). Computing the aspect graph for line drawings of polyhedral objects. In: Proceedings: 1988 IEEE International Conference on Robotics and Automation (pp. 1560–1566). IEEE.
26. Hromkovič, J., (1999). Communication complexity and lower bounds on multilective computations. RAIRO-Theoretical Informatics and Applications, 33(2), 193–212.
27. Jeyalakshmi, M., & Meena, N., (2021). Open support of some special types of graphs under addition. World Scientific News, 156, 130–146.
28. Kennedy, G. J., Kenway, G. K. W., & Martins, J. R. R. A., (2014). High aspect ratio wing design: Optimal aerostructural tradeoffs for the next generation of materials. In: Proceedings of the AIAA Science and Technology Forum and Exposition (SciTech), (pp. 2014–0596). National Harbor, MD.
29. Kimoto, K., Endo, K., Iwashita, S., & Fujiwara, M., (2005). The application of PDA as mobile computing system on construction management. Automation in Construction, 14(4), 500–511.
30. Kozae, A. M., El Atik, A. A., Elrokh, A., & Atef, M., (2019). New types of graphs induced by topological spaces. Journal of Intelligent & Fuzzy Systems, 36(6), 5125–5134.
31. Kuchment, P., (2008). Analysis on Graphs and its Applications. In: Proc. Symp. Pure. Math (Vol. 77, No. 291, pp. 0802–3442).
32. Liotta, G., & Battista, G. D., (1995). Computing proximity drawings of trees in the three-dimensional space. In: Workshop on Algorithms and Data Structures (pp. 239–250). Springer, Berlin, Heidelberg.
33. Maass, W., (1997). Bounds for the computational power and learning complexity of analog neural nets. SIAM Journal on Computing, 26(3), 708–732.
34. Mahmoodi, M. R., Vincent, A. F., Nili, H., & Strukov, D. B., (2020). Intrinsic bounds for computing precision in memristor-based vector-by-matrix multipliers. IEEE Transactions on Nanotechnology, 19, 429–435.
35. Pardalos, P. M., & Kovoor, N., (1990). An algorithm for a singly constrained class of quadratic programs subject to upper and lower bounds. Mathematical Programming, 46(1), 321–328.
36. Pirandola, S., & Lloyd, S., (2008). Computable bounds for the discrimination of Gaussian states. Physical Review A, 78(1), 012331.
37. Pizzarelli, M., Nasuti, F., & Onofri, M., (2013). Trade-off analysis of high-aspect-ratio-cooling-channels for rocket engines. International Journal of Heat and Fluid Flow, 44, 458–467.

38. Pratt, V. R., & Yao, F. F., (1973). On lower bounds for computing the i-th largest element. In: 14th Annual Symposium on Switching and Automata Theory (swat 1973) (pp. 70–81). IEEE Computer Society.
39. Rebolj, D., & Menzel, K., (2004). Mobile computing in construction. J. Inf. Technol. Constr., 9, 281–283.
40. Samee, M., Hassan, A., Irfan, M. T., & Rahman, M., (2008). Computing β-drawings of 2-outerplanar graphs in linear time. In: International Workshop on Algorithms and Computation (pp. 81–87). Springer, Berlin, Heidelberg.
41. Tamassia, R., (1996). Strategic directions in computational geometry. ACM Computing Surveys (CSUR), 28(4), 591–606.
42. Tamassia, R., (2000). Graph drawing. In: Handbook of Computational Geometry (pp. 937–971). North-Holland.
43. Traub, J. F., & Woźniakowski, H., (1976). Strict lower and upper bounds on iterative computational complexity. In: Analytic Computational Complexity (pp. 15–34). Academic Press.
44. Zhongke, W., Feng, L., Soon, S. H., & Yun, C. K., (2004). Evaluation of difference bounds for computing rational Bézier curves and surfaces. Computers & Graphics, 28(4), 551–558.

INDEX

D

E

F